IN SEARCH OF SUSTAINABILITY

IN SEARCH OF SUSTAINABILITY

Edited by
Jenny Goldie, Bob Douglas and Bryan Furnass

CSIRO
PUBLISHING

National Library of Australia Cataloguing-in-Publication entry

In search of sustainability.
Includes index.
ISBN 0 643 09062 2.

1. Sustainable development – Australia.
2. Environmental policy – Australia.
3. Economic development – Environmental aspects – Australia.
I. Goldie, Jenny. II. Douglas, Robert Matheson, 1936– .
III. Furnass, Bryan. IV. Australia 21. V. Nature and Society Forum.
VI. Sustainable Population Australia. VII. ISOS Online Conference (2003 : Canberra, A.C.T.).

338.927

Available from
CSIRO PUBLISHING
150 Oxford Street (PO Box 1139)
Collingwood VIC 3066
Australia

Telephone:	+61 3 9662 7666
Local call:	1300 788 000 (Australia only)
Fax:	+61 3 9662 7555
Email:	publishing.sales@csiro.au
Web site:	www.publish.csiro.au

Cover
Wind turbine in Geelong, Victoria. Photo by Lisa Schneider.

Set in 10.5/13 Adobe Minion and Stone Sans
Cover and text design by James Kelly
Typeset by Desktop Concepts P/L, Melbourne
Printed in Australia by BPA Print Group

Contents

Acknowledgements

The editors are grateful to Professor Frank Fenner for his generous donation towards the production costs of this publication.

We also wish to thank the many participants who made valuable contributions to the sustainability debate during the 2003 nine-month Internet conference 'In Search of Sustainability' and in the culminating face-to-face conference held in the Shine Dome of the Australian Academy of Science on 14 November 2003. Their contributions can be found on the conference website www.isosconference.org.au.

Preface

In two and a half centuries of industrial civilisation there has been a tenfold growth in global population, great increases in economic activity and food production, a markedly enhanced life expectancy, and generally huge achievements in material prosperity, employment prospects and human well-being.

But the price of this progress has been widespread land and water degradation, loss of species biodiversity and inequalities in resource distribution both between and within industrialised and developing countries. The excessive combustion of fossil fuels which drive agriculture, city construction and maintenance, manufacturing industries and transport is also contributing to significant climate change which, in turn, is threatening the very ecosystems on which all humans depend for life.

As human numbers have increased and technology improved, terror and war have escalated to new heights of inter-racial and inter-religious barbarity.

It has been estimated that humans now appropriate over 40 per cent of the planet's photosynthetic activity and that the biosphere passed its regenerative and absorptive capacity a quarter of a century ago. Depletion of nonrenewable resources such as phosphorus, fossil fuels and the rich biodiversity of old-growth forests and marine ecosystems has reached a critical stage. Governments define progress in terms of economic growth. Yet if ecological costs exceed economic benefits, growth is making us poorer rather than richer. A transition to sustainability is a no less important chapter in human history than the agricultural and industrial transitions, whose excesses have contributed to the environmental crisis which we now confront.

In 1987 the World Commission on Environment and Development published a report 'Our Common Future' which defined sustainable development as 'that which meets the needs of the present without compromising the ability of future generations'.

Since that time, the word 'sustainability' has been widely used by environmentalists, government agencies and politicians. The word has become so extensively and rhetorically used through a decade of further decline in real sustainability, that it is in danger of becoming meaningless.

But if we are not sustainable as a human species, our children or their children could die very young. *Homo sapiens* could quite quickly become extinct. It is as simple and serious as that.

In 2003 we convened a nine-month Internet conference 'In Search of Sustainability', and invited a number of leading Australian thinkers and researchers to provoke discussion on nine themes that bear on the concept. The themes were Water, Health and Well-being, Land Use and Natural Ecosystems, Economic Systems, Equity and Peace, Energy, Climate Change, Labour Force and Work and Urban Planning and Transport.

The nine keynote papers provoked 76 others, and considerable cyber-discussion on these nine themes. At a face-to-face meeting in Canberra in November 2003, two hundred people from many walks of Australian life met to consider a communiqué that contained 34 recommendations for action. These recommendations are contained in the appendix to this volume and the full contents of the internet conference and the communiqué are accessible at www.isosconference.org.au.

After the conference, we invited 11 of the conference speakers to prepare a chapter, drawing together their considered views in the context of the conference discussions. This book is the result. The authors are national leaders in their subject areas and we hope that their arguments will be widely considered and discussed in the Australian community. Our own concern at the urgency of the issue is reflected in Chapter 1.

One of the suggestions at the Canberra conference was for a Ministry of Sustainability to become our most senior government ministry, taking precedence over Treasury. Another proposal which received strong support was for a far-reaching enquiry into sustainability of equal significance and impact to the Hilmer Enquiry which has resulted in competition policy becoming embedded in all aspects of national and state government and corporate policy.

We can no longer regard sustainability as a luxury. The consensus is building internationally that we do not have much time left if we are to leave a habitable planet for our descendants.

Australia has been a 'lucky country'. Many believe we could join the world leaders in the race towards a sustainable global future. This book seeks to chart the way we could begin to do that.

Jenny Goldie, Sustainable Population Australia
Bob Douglas, Australia 21
Bryan Furnass, Nature and Society Forum

About the Contributors

Professor Andrew Blakers is the Foundation Director of the Centre for Sustainable Energy Systems at the Australian National University. He has a physics degree from ANU and a PhD from the University of New South Wales. He was a Humboldt Fellow at the Max Planck Institute for Solid State Physics in Stuttgart and has held ARC QEII and Senior Research Fellowships. His research interests are in the areas of photovoltaics and sustainable energy systems. http://solar.anu.edu.au

Associate Professor John Burgess is Director, Employment Studies Centre, and Assistant Dean for Research, Faculty of Business and Law, University of Newcastle. His research interests include gender and work, workplace change, employment restructuring and labour market policy.

Dr Colin Butler, a co-founder of BODHI, a Third World development organisation, has visited the Third World many times. He is currently writing *The Human Titanic*, a book about global inequality and sustainability. He is a coordinating lead author for the chapter on human well-being in the scenarios working group of the Millennium Ecosystem Assessment and is a research fellow at the National Centre for Epidemiology and Population Health at the ANU.

Associate Professor Julia Connell is Deputy Director, Employment Studies Centre, and Assistant Dean (International), Faculty of Business and Law, at the University of Newcastle, Australia. Her research interests include change management, organisational effectiveness, various aspects of temporary work, call centres and labour turnover.

Professor Peter Cullen AO, FTSE, is one of the Wentworth Group of Concerned Scientists who seek to raise public awareness of natural resource issues facing Australia. He is Chair of the Victorian Water Trust Advisory Council, and a Director of both Land and Water Australia and Landcare Australia Ltd. He retired in 2002 after 10 years as Chief Executive of the CRC for Freshwater Ecology at the University of Canberra. He is a Visiting Fellow at CSIRO Land and Water. Professor Cullen was awarded the Prime Minister's Prize for Environmentalist of the Year in 2001 for his work on the National Action Plan for Salinity and Water Quality. He is a Member of the International Water Academy.

Richard Denniss is Deputy Director of The Australia Institute, a Canberra-based public interest think-tank. Prior to his current position Richard was employed as Chief of Staff to the then leader of the Australian Democrats, Senator Natasha Stott Despoja. Richard was also employed as an Economics Lecturer at the University of Newcastle. Richard's current research interests include the economics of regulation, particularly in relation to the development of environmental policy.

Emeritus Professor Bob Douglas AO was the Foundation Director at the National Centre for Epidemiology and Population Health at The Australian National University from 1989–2001 where he remains a Visiting Fellow. He is Chair of The Board of Australia 21, a non-profit organisation that is committed to developing networks of understanding and influence on a number of the large topics that challenge Australia's future. www.australia21.org.au

Dr Bryan Furnass AM is a retired physician who was Foundation Director of the University Health Service at the Australian National University from 1966–91, where he pursued his interests in preventive medicine and health promotion. Since retirement from clinical practice, as a member of Nature and Society Forum, he has developed an amateur concern with the health and sustainability of the natural environment, believing that both medical and environmental science would benefit from a more holistic and integrative approach.

Ms Jenny Goldie was National Director of Sustainable Population Australia Inc from 2000–2004, having helped establish the organisation (as Australians for an Ecologically Sustainable Population Inc.) in Canberra in 1988. She is a former science teacher and science communicator. She is primarily interested in the interrelationship between population, development and environment.

Dr Clive Hamilton is Executive Director of The Australia Institute, a Canberra-based public interest think-tank. His research work at the Institute has emphasised climate change policy, measures of well-being (especially the Institute's Genuine Progress Indicator), privatisation and taxation issues. Clive is also a Visiting Fellow in the Graduate Program in Public Policy at the ANU and an Adjunct Professor at the University of Technology, Sydney. He has published widely, including six books.

Emeritus Professor Ian Lowe AO has been with the School of Science at Griffith University since 1980. He directed the Commission for the Future in 1988 and chaired the advisory council which produced the first national report on the state of the environment in 1996. He is currently involved in international work on 'sustainability science' and the Great Transition Initiative.

Professor Tony McMichael, epidemiologist, directs the National Centre for Epidemiology and Population Health at the ANU. His research interests span various environmental and social influences on health, including the health impacts of climate change and other global environmental changes. His most recent book is *Human Frontiers, Environments and Disease: Past Patterns, Uncertain Futures*, Cambridge University Press (2001). He recently edited *Climate Change and Human Health: Risks and Responses*, published by WHO (2003).

Professor Peter Newman is Professor of City Policy and Director of the Institute for Sustainability and Technology Policy, at Murdoch University. He is Chair of the Western Australian Sustainability Roundtable advising the Premier on how to implement their Sustainability Strategy. Peter's book with Jeff Kenworthy *Sustainability and Cities: Overcoming Automobile Dependence* was launched in the White House in 1999 and his 2001 co-authored book is called *Back on Track: Rethinking Australian and New Zealand Transport*. He is a Sustainability Commissioner in NSW.

Dr Graeme Pearman AM FAA was the Chief of CSIRO Atmospheric Research 1992–2002. He recently established the CSIRO CLIMATE Program involving research from 13 Divisions. He now operates his own consultancy company and works with Monash University establishing new endeavours in sustainability science. He chairs the Australian Academy of Science National Committee for Sustainability and the Joint Academies Committee for Sustainability. Internationally he is co-Chair of START International (Washington DC) and past co-Chair of the Asia Pacific Network (Kobi) science steering committees, two science capacity building organisations for global environmental change research.

Dr Denis Saunders is a conservation biologist and landscape ecologist. He is a former CSIRO Chief Research Scientist, and is a member of the newly formed Australian Heritage Council. He edits two international scientific journals *Biological Conservation* and *Landscape and Urban Planning*. He is a member of WWF's Executive Council, a Director of Earthwatch Institute Australia, a member of the Council of Birds Australia, and of the Wentworth Group of Concerned Scientists.

Dr John Williams was raised on a grazing property on the southern tablelands of New South Wales, and after graduating from the University of Sydney conducted research in Canada, USA and the South Pacific, before joining CSIRO in Townsville, where he spent 16 years working on landscape hydrology in the Australian tropics. He was appointed Deputy Chief of CSIRO Land and Water at its inception in 1997 and Chief in September 2001. Following retirement from CSIRO in 2004 he accepted a position as Chief Scientist for the newly formed Department of Infrastructure Planning and Natural Resources in NSW. He is a member of the Wentworth Group, a group of 11 independent scientists who have contributed to new policies on water and landscape management.

1 – An urgent need to change direction

Jenny Goldie, Bob Douglas and Bryan Furnass

The Earth's life support systems are in peril. As a species, we are not living sustainably and are accelerating towards collapse of the natural capital on which human society and its economy depends. Sustainability refers to the capacity of the biosphere to provide for the full range of human concerns in the long term. In Australia we are faring well economically but both our social and environmental systems are showing evidence of serious damage. At a global level, inequity and inequality of opportunity lie at the heart of global instability, terror and wars. If we are to survive and prosper as a species, there will need to be a drastic change in conventional values, economic structures and social arrangements. We must also plan our own national development in the context of the 'globalised' interdependence of all human populations on each other and on the natural world. We must now make the transition towards sustainability. Australia could be at the leading edge of this endeavour.

Human activities are now significantly affecting the planet and how it functions. Not only the magnitude, but also the rate of human-driven change, is alarming. Atmospheric carbon dioxide is rising, global mean temperature is increasing, huge areas of humid tropical forest are cleared each year, and biodiversity losses, currently driven by habitat loss, will be exacerbated by further global climate change. Already two billion people suffer from severe water stress, but by 2025 it may be four billion. There are significant risks of rapid and irreversible environmental changes, to which it would be difficult to adapt and which could have severe economic and societal consequences.

This warning was issued in early 2004 by a team of four including Margot Wallstrom, European Commissioner for the Environment and

Australian Will Steffen, director of the international Geosphere–Biosphere Program (Wallstrom 2004). It came shortly after the equally alarming prediction in the journal *Nature* (Thomas 2004) that a quarter of species will be extinct by mid-century under a mid-range global warming scenario. Even at the lowest range of climate change, 18 per cent would be extinct. The authors stressed that these '... estimates show the importance of rapid implementation of technologies to decrease greenhouse gas emissions and strategies for carbon sequestration'.

In 2002 an earlier wake-up call had come from a team of scientists led by Mathis Wackernagel in a paper published by the US National Academy of Sciences (Wackernagel 2002). Humanity's collective demands surpassed the earth's regenerative capacity around 1980. Every year since, we further exceed that capacity by another 1 per cent, Wackernagel said.

The evidence suggests that progress in the human economy has been at the expense of the earth's natural assets. We are accelerating towards collapse of the natural capital on which human society and its economy depends. We are not living sustainably.

There were even earlier warnings. The issue of sustainability first emerged with the publication of *The Limits to Growth* in 1972 (Meadows 1972), the same year as the first Earth Summit. The book concluded that the finite nature of the natural environment meant that, in terms of material throughput, the world economic system could not expand indefinitely. It said, however, that if actions were taken to modify current trends, the world economic system could move into a configuration that would be 'sustainable far into the future'. Because of the environmental limits on continuing economic growth, meeting the needs of the poor would have to come through major redistribution of wealth and income from rich to poor both between and within nations.

Subsequently in 1987, Gro Harlem Brundtland, Prime Minister of Norway, as head of the World Commission on Environment and Development, introduced a concept of 'sustainable development' in the book *Our Common Future* (Brundtland 1987). Widely known as 'The Brundtland Report' it defined sustainable development as that which:

> ***... seeks to meet the needs and aspirations of the present without compromising the ability to meet those in the future.***

In contrast to *The Limits to Growth*, which required an ultimate cessation of economic growth, the Brundtland Report stated that the problems of poverty and underdevelopment could be solved only by a 'new era of growth in which developing countries play a large role and reap large benefits'. Industrialised nations could and should continue their own economic growth of 3–4 per cent, and these nations would be environmentally sustainable if this growth was to come from fewer material and energy-intensive activities and improved efficiency in using materials and energy.

The definition of sustainability

In this book we use the term 'sustainability' to refer to the capacity of human systems to provide for the full range of human concerns in the long term. Sustainability, when applied to humans, refers both to long-term survival of our species and the quality of our lives.

The warnings referred to above suggest that we live in an unsustainable world and that our current Australian way of life is not sustainable in the context of the changes that are taking place across the planet.

Australians are increasingly expressing a sense of unease about the legacy we are developing for our children. And this is reflected in a widespread public acceptance of the need for environmental conservation.

But the full ramifications of a sustainable society are not yet being seriously debated in the broader Australian community. Some government agencies are beginning to articulate the concept in all its complexity. The Western Australian government has recently established a series of principles and processes as part of its search for a sustainable future. These are described in the accompanying table (Table 1.1).

These principles highlight the interconnectedness of social, environmental and economic activity. They also underline the reasons why governments are experiencing difficulty with the task.

A truly sustainable society will require a profound change in mindset and a reorientation of the values of our national culture. The principles do not lend themselves to simplistic slogans. Everything in society is connected to everything else and sustainability permeates it all.

A revitalisation of national thinking about the sustainability challenge is taking place and Peter Newman, who is working with the Western Australian Government on implementation of its policy, argues that sustainability:

Table 1.1 Principles of sustainability

Foundation principles
Long-term economic health Sustainability recognises the needs of current and future generations for long-term economic health, innovation, diversity and productivity of the earth.
Equity and human rights Sustainability recognises that an environment needs to be created where all people can express their full potential and lead productive lives and that significant gaps in sufficiency, safety, and opportunity endanger the earth.
Biodiversity and ecological integrity Sustainability recognises that all life has intrinsic value, is interconnected and that biodiversity and ecological integrity are part of the irreplaceable life support systems upon which the Earth depends.
Settlement efficiency and quality of life Sustainability recognises that settlements need to reduce their ecological footprint (that is less material and energy demands and a reduction in waste), while they simultaneously improve their quality of life (health, housing, employment, community, and so on).
Community, regions, 'sense of place' and heritage Sustainability recognises the significance and diversity of community and regions for the management of the earth, and the critical importance of 'sense of place' and heritage (buildings, townscapes, landscapes and culture) in any plans for the future.
Net benefit from development Sustainability means that all development, and particularly development involving extraction of non-renewable resources, should strive to provide net environmental, social and economic benefit for future generations.
Common good from planning Sustainability recognises that planning for the common good requires equitable distribution of public resources (like air, water and open space) so that ecosystem functions are maintained and so that a shared resource is available to all.
Process principles
Integration of the triple bottom line Sustainability requires that economic, social and environmental factors be integrated by simultaneous application of these principles, seeking mutually supportive benefits with minimal trade-offs.
Accountability, transparency and engagement Sustainability recognises that people should have access to information on sustainability issues, that institutions should have triple bottom line accountability, that regular sustainability audits of programs and policies should be conducted, and that public engagement lies at the heart of all sustainability principles.

Table 1.1 Principles of sustainability (continued)

Precaution Sustainability requires caution, avoiding poorly understood risks of serious or irreversible damage to environmental, economic or social capital, designing for surprise and managing for adaptation.
Hope, vision, symbolic and iterative change Sustainability recognises that applying these principles as part of a broad strategic vision for the Earth can generate hope in the future, and thus it will involve symbolic change that is part of many successive steps over generations.
Source: Western Australian Government (2003).

- is the agenda of our age
- is a practical and useful concept that can integrate across professions and disciplines
- can provide the vision we need to draw together government, the private sector community and academics to help solve our many deep-seated problems
- is a politically useful concept which can provide politicians with the means of expressing leadership at a time of global fear and backward-looking
- is therefore a great source of hope (Western Australian Government 2003).

Australia and ecologically sustainable development

Australia responded in 1992 to the Bruntland Report by developing a National Strategy for Ecologically Sustainable Development (NSESD 1992). It commits all Australian Governments to:

- enhance individual and community well-being and welfare by following a path of economic development that safeguards the welfare of future generations
- provide for equity within and between nations
- protect biological biodiversity and maintain ecological processes and life support.

Twelve years later, the Australian economy has expanded – national incomes and wealth have risen. Many now wonder, however, whether our unqualified commitment to economic growth may be a threat to the sustainability of our society. By almost all criteria, the Australian environment continues to degrade. Given that the natural environment provides the ecological and material basis for human existence, we may be compromising the welfare of future generations.

It is noteworthy that the ESD objectives use a combination of economic, social and ecological criteria. Climate change, energy, water and the health of natural ecosystems loom as major environmental issues. Human health and well-being, equity and peace, and labour force and work are significant social issues impacting on sustainability. Land use, economic systems, transport and urban design are major social and ecological issues. These will all be addressed in the following chapters.

The Australian Bureau of Statistics (ABS) has recognised the interrelationships between the economic, social and environmental aspects of life. Realising the need to go beyond Gross Domestic Product (GDP) as a measure of human progress, it produced a suite of indicators in its 2002 publication *Measuring Australia's Progress* (ABS 2002). These suggest progress in some areas of Australian life and regress in others. While 'progress' is by no means synonymous with 'sustainability', it is useful to examine these indicators to help us along the road to sustainability.

On almost every count, Australia is faring well economically. National wealth grew during the 1990s and real wealth per person grew modestly, by 1 per cent a year. Real income per head grew more strongly. Incomes of the less well off (though not the poorest) grew by 5 per cent in the three years to 1997–98, as did those of better-off groups.

Australia's natural environment

But how is Australia faring environmentally? According to *Australia State of the Environment 2001* (Australian SoE Committee 2001), there are some environmental good news stories. For instance, urban air quality has generally improved; sulphur dioxide emissions have decreased substantially; fewer turtles are being caught as by-catch; there is better storm-water management; another 17.6 million hectares of marine areas are now protected since 1996; biodiversity can be better protected thanks to passage of the *Environmental Protection and Biodiversity Conservation Act* 1999 (EPBC Act); energy efficiency in residences has improved; and domestic water use per capita declined for most urban centres during the 1990s because of water pricing, consumer education, use of water-saving appliances and higher residential densities.

Set against these improvements, however, the *State of the Environment* report listed the high per capita greenhouse gas emissions; increasing land surface temperatures; loss of coastal habitat through the encroachment of human settlements; pressures on our coral reefs from the downstream effects of land use; large nutrient loads of nitrogen and phosphorus being discharged to coastal and estuarine waters; the net loss of vegetative cover from broadacre clearing; turbidity in waterways from soil erosion; large areas of acidic soils; continuing deterioration of the health of surface and groundwaters; an overall increase in water use and extraction for irrigation; increasing algal blooms in waterways; persistence of the key threats to biodiversity such as salinity, land-clearing and exotic invasive species; consumption outpacing population growth in electricity generation and transport usage; and environmental noise and its effects on residents is increasing from high residential densities and volume of traffic.

Australia currently has 5.1 per cent of the world's land area but only 0.3 per cent of its human population, largely because 70 per cent of the continent is arid. Even with one of the lowest population densities in the world, our environment is being significantly degraded by human activity.

Water is perhaps our most precious resource. Australia is the driest inhabited continent, characterised by variable climatic conditions and high levels of evaporation. There is relatively little run-off, so Australian rivers have low and variable flows and carry high levels of nutrient and sediment as a consequence of agricultural land use in the catchment areas (ABS 2003). Salinity and algal blooms also threaten Australian rivers and wetlands.

Meanwhile, Australia's unique biodiversity is threatened by a number of other factors, the major one being clearance of native vegetation. Since European settlement, around 100 million hectares have been cleared, mostly for agriculture. Over half a million hectares of native vegetation were cleared in 2000, a rate exceeded by only four other countries. Clearing of native vegetation not only destroys plants and habitats for animals, it helps invasive species to spread which then compete with native wildlife. These include foxes, cats, rabbits, goats and dieback fungus.

Australian soils are old, shallow and susceptible to degradation. Dryland salinity, caused by rising watertables bringing salts to the surface, is affecting 20 000 farms across two million hectares, much of it in the Western Australian wheat belt. Another 15 million hectares are at risk, including large areas of natural vegetation and farms of the Murray–Darling Basin, the nation's breadbasket. Salinity not only degrades agricultural lands, rivers and wetlands, it affects houses, roads and water supply infrastructure.

Perhaps even worse than salinity, acidification degrades our soils. Certainly it has cost more, at over $1 billion in 2000 compared to $187 million in lost agricultural production from salinity. These changes, as well as erosion, altered fire and grazing regimes, and pests and weeds all contribute to declining health of farms and rangelands.

It is climate change, however, that looms as perhaps the greatest environmental threat. Since 1910, Australian average surface temperature has increased by 0.76°C. Temperatures may rise anywhere from 1.4°C to 5.8°C this century from 1990 levels. Global warming poses massive risks for Australian biodiversity, in some cases even exceeding the average global losses. At higher estimates of warming, for instance, Queensland forests could lose about 80 per cent of their birds, mammals, reptiles and frogs by 2050. A team from the Rainforest Co-operative Research Centre says that while there have been climatic changes in the past affecting biodiversity, this time the effect will be worse (Krockenberger et al. 2003). Climate change today, they say, differs from past variability in two ways. First, the rate of change recorded in the late 20th century, and predicted to continue, is considered by many scientists to be unprecedented in the past 10 000 years. Second, many of the earth's ecosystems are already stressed by other human impacts, such as land clearing and the consequent fragmentation of natural vegetation.

Fossil fuel combustion is a major contributor to Australia's greenhouse gas emissions. Our total consumption of energy over the 1990s grew twice as fast as population growth. We are thus contributing to the very global warming that may ultimately devastate our biodiversity. Greenhouse gas emissions rose 17 per cent between 1990 and 1999 despite our signing (although the Federal Government now refuses to ratify) the Kyoto Protocol in which we committed ourselves to limit emissions to 8 per cent over 1990 levels by 2008–12.

The social environment in Australia

And what of the social environment? Australians' health is improving in some ways – children born in 1999 are likely to live three more years on average than children born in 1990 – and Australians are now amongst the longest-lived people in the world. But Australian of the Year for 2003, Fiona Stanley, has drawn national attention to the fact that many indicators of developmental health and well-being are showing adverse trends among children and adolescents in Australia. Rising rates are being observed for low birth weight, neuro-developmental disorders, asthma, type 1 diabetes, inflammatory bowel disease, autism, mental health morbidities, child abuse and neglect, adolescent suicide, obesity, eating disorders, learning disabilities, behavioural disorders, aggressive behaviours and violence, school dropout and truancy, juvenile crime, illicit drug and alcohol use, and teenage births (Stanley 2002).

Heart disease, cancer and strokes, which had replaced infectious diseases as the main cause of death, have declined. There are huge differences in health within the population, however, and Indigenous Australians have a life expectancy 20 years less than the national average.

Despite gains in life expectancy amongst non-Indigenous Australians there is no cause for complacency in regard to either infectious or non-infectious diseases.

The widespread and inappropriate use of broad-spectrum antibiotics such as vancomycin-type drugs as 'growth promoters' in animals reared for food and their profligate non-specific prescription for humans has led to the emergence of multi-resistant strains of the bacteria *Staphylococcus aureus* and *Escherichia coli (E. coli).* If these bacteria become more prevalent it will be dangerous to be admitted to hospital and we may return to the pre-antibiotic days in which staphylococcal infections carried a mortality rate of 30 per cent or more (Collignon 2002). We may also be witnessing changes in the epidemiology of respiratory viruses. One consequence of the growth in numbers and density of the human population may be epidemics of new mutants of highly transmissible viruses. Recent years have witnessed the emergence of 'new' viral diseases such as SARS (severe acute respiratory syndrome) and the bird influenza virus in Asia; the latter carries the small but potentially dangerous chance of gene re-assortment with human influenza virus. With the instantaneous spread by air transport of infected cases across countries and continents, such diseases have potential to be profoundly lethal to populations which have no immunity to them, and could seriously disrupt the sustainability of mass international air travel. (Witness the effects of the swine flu epidemic in 1919 when population densities were much less than they are now.)

Global warming is likely to increase the spread of vector-borne infectious diseases such as malaria, dengue, Ross River virus and Japanese encephalitis to higher altitudes and wider latitudes, including to the Northern Territory of Australia and North Queensland.

In regard to non-infectious diseases, Australia is one of the affluent countries with a rising epidemic of the so-called metabolic syndrome of obesity, type 2 diabetes (incidence has doubled in 20 years), and associated cardiovascular disorders. Paradoxically, this syndrome is commoner among Indigenous than among non-Indigenous communities. Surveys have shown that physical inactivity (from excessive use of television, computers and car travel) by our sedentary population may be as much to blame as excessive consumption of energy-rich well-advertised foods. The epidemic calls for radical educational and lifestyle changes, particularly among the younger generation.

At the other end of life's spectrum, our therapeutic success at keeping people alive well into their eighties has led to a rising incidence of chronic disorders such as arthritis and various forms of dementia, including Alzheimer's disease. It is now becoming recognised that half our medical resources are devoted to people in the last year of life, producing an ongoing shortage of hospital and nursing home beds. Medical practice in future must become increasingly focused on improving the quality of life rather than simply increasing its quantity. This raises profound ethical questions about whether it is justifiable or humane to keep people alive against their wishes, when the prospect of quality living is close to zero. There is already some vigorous debate about whether there is a right to die – which is no less persuasive than the right to live – including the relative merits or demerits of passive versus active voluntary euthanasia, given appropriate community safeguards.

With respect to other social indicators, some are improving such as education and training. Overall, the proportion of people with a vocational or higher education qualification continues to rise, though as with health, Indigenous participation in education and training and their levels of attainment remain well below that of the total population.

Unemployment has declined since the early 1990s but remains higher than was common in the 1960s and 1970s. Once Australia was one of the most egalitarian countries in the world thanks to a regulated labour market assuring the average full-time worker minimum wages, sick leave and holidays. A minimum wage – for a male at least – was meant to support a spouse and two children. But now almost half the workforce is employed in jobs that are casual, part-time or on fixed contracts. Over a million workers are unemployed, underemployed or in hidden employment. Many workers are working longer hours without overtime or their jobs have intensified in other ways. All these are impacting on family life. Meanwhile, a degree of panic has set in about the ageing population and a future 'lack of workers' despite Australia's dependency ratio (workers to dependents) being healthy for another two or three decades.

This change in the nature of the workforce affects other things. Income inequality, for instance, is rising despite an increase in income across most groups in percentage terms. Certainly, there is a perception of inequality. A 2003 poll conducted by Roy Morgan Research of 19 000 Australians found that 88 per cent agreed with the statement: 'I think the gap between rich and poor is growing'. The Australian Council of Social Service noted in January 2004 that the real taxable incomes of the bottom 5 per cent had actually fallen by nearly 7 per cent to $7648 in the five years to 2000–01, while earnings of the top 5 per cent rose by over 31 per cent to $150 820, which is about 20 times the income of the poorest age group.

Housing costs, especially in the major cities where high population growth, low interest rates and other factors have led to significant inflation, exacerbates this inequality. Housing affordability in 2003 was the worst in 13 years, prompting the Government to request the Productivity Commission to undertake an inquiry. Not only are young people increasingly unable to enter the market, more and more people are rendered homeless because of the lack of adequate low cost housing. In 1996, the Australian Census recorded 105 000 people as officially homeless. Since then, community organisations providing crisis accommodation and support services recorded an increase from about 15 000 to 16 000 seeking help each day, though this represents only a fraction of the number of homeless people.

The second ESD objective – to provide for equity within and between nations – is nowhere being reached, in fact, inequity is increasing nationally and internationally. As already noted, the gap between rich and poor has widened in Australia. Globally, despite the spectacular transition of some poor nations out of poverty, others remain 'demographically entrapped'. Their populations have exceeded the carrying capacity of their ecosystems, they are unable to migrate and their economies do not generate enough exports to buy food and other essentials on the world market. Meanwhile Australia's commitment to overseas aid has dropped to 0.24 per cent of GDP, way below the UN-recommended 0.7 per cent. The debts of many developing countries remain an unsustainable financial burden yet few richer countries have offered to cancel such debts.

Global non-sustainability

Yet it can be argued that inequity and inequality of opportunity lie at the heart of global instability, terror and wars. As populations come up against the limits of their resource base, resentment towards the better off may lead to riots and anarchy. When it crosses national boundaries, it becomes terrorism or war. It has often been said that the wars of the 21st century will be fought over water. This is not surprising, given more than three hundred river systems cross national boundaries, and access to water means irrigation water and food. Some major rivers of the world, for example, the Nile, the Ganges and the Tigris–Euphrates, pass through a number of countries with rapidly expanding populations, all anxious to feed their own people and pull themselves out of poverty. But wars are being fought over access to other resources as well, notably oil. Preparations for war divert a country's resources away from health, education and welfare and exacerbate inequity.

Discussions on sustainability too often have been distorted by the assumption that economic growth is the only sign of a healthy economy. Yet economic growth that turns human labour, ecosystems and material resources into waste and pollution, is clearly unsustainable. Economists forget that the human economy depends absolutely on the materials that nature provides. *The Limits to Growth* was right: given that resources are finite, we have to move away from the dominant paradigm of unending economic growth towards a dynamic, steady state economy and stay within ecological and material limits.

Oil experts are warning that sometime in the next decade, possibly around 2010, global oil supplies will peak. As demand starts to exceed supply, prices will inevitably rise. This has huge implications for transport and the design of cities. Unless cities reduce their oil consumption they will not be sustainable for long. Some argue for more compact, densely populated cities, with more accessibility by walking, cycling or public transport, while others argue for reduced densities but with technological changes to cars. A combination of both may be the answer to the controversy, with the compact areas having more intense economic activity and the low-density areas having green activities such as urban agriculture. City buildings could be transformed by gardens on balconies and roofs.

The decline of the age of oil will require a transition to other transport fuels, first to gas and then to hydrogen. Rail, already cheaper than road transport, may become the predominant transport mode as it can be run on electricity. Even with current technology, it seems feasible to produce significant amounts of electrical power with solar-based energy. Achieving government commitment to adequate research in renewable energy, however, may be more difficult than technological advance itself.

What should be done?

So how do we forge a new sustainability paradigm?

We stated earlier that the term *sustainability* refers to the capacity of human systems to provide for the full range of human concerns over the long term and that sustainability, when applied to humans, refers both to long-term survival of the species and the quality of their lives.

Global sustainability will require that we recognise and act upon the gross inequalities of access to natural resources and to health care between industrialised and developing countries.

In moving towards a sustainable future, we will need to ensure that we avoid eroding the natural resource base, provide substantial global equality of opportunity, and keep the size of the human population within ecological limits.

In addition, if in future we wish to address the needs of all Australians, all will need to feel connected to society and have a sense of their role and place in it. In this, we cannot avoid issues of spirituality. At the broadest level, spirituality can be thought of as that part of their lives that provides individuals with a sense of 'connectedness', meaning and purpose. In the view of many, the global epidemic of drug dependency and the Australian suicide epidemic are related to an increasingly unmet human need to feel connected both to the natural world and to the human family.

Now that we belong to a globalised world, it no longer makes sense to think of sustainability only at the national level. There are however, major national issues that must be addressed at the same time as Australia plays a humane and responsible role in the world community.

Values change needed

According to the Global Scenario Group, the transition to sustainability is going to require a drastic change in conventional values, economic structures and social arrangements (Raskin 2002). The first wave of sustainability activity in progress since the Earth Summit of 1992 is insufficient to alter alarming global developments, it says. It argues that four major agents of change acting synergistically could drive the paradigm. Three are global actors -- intergovernmental organisations, transnational organisations and spiritual communities. The fourth is less tangible but is critical: *wide public awareness of the need for change and the spread of values that underscore quality of life, human solidarity and environmental sustainability.* As the Global Scenario Group says:

> ***In the critical years ahead, if destabilizing social, political and environmental stresses are addressed, the dream of a culturally rich, inclusive and sustainable world civilization becomes plausible. If they are not, the nightmare of an impoverished, mean and destructive future looms. The rapidity of the planetary transition increases the urgency for vision and action lest we cross thresholds that irreversibly reduce options – a climate discontinuity locking-in to unsustainable technological choices and the loss of cultural and biological diversity. Postponing the rectification of how we live together on this planet could foreclose the opportunity for a Great Transition (Raskin 2002).***

This is what the rest of this book is all about.

References

Australian Bureau of Statistics. (2002). 'Measuring Australia's Progress', Catalogue 1370.0, Commonwealth of Australia.

Australian Bureau of Statistics. (2003). 'Environment by Numbers' Commonwealth of Australia.

Australian State of the Environment Committee. (2001). *Australia State of the Environment 2001.* Independent Report to the Commonwealth Minister for the Environment and Heritage. CSIRO Publishing on behalf of the Department of the Environment and Heritage, Canberra.

Brundtland, G.H. (1987). *Our Common Future.* Oxford University Press, Oxford.

Collignon, P. (2000). Antibiotics. In *Bad Bugs – People and Infectious Diseases.* (Eds B. Furnass and S. Haygarth). pp. 43–47. Nature and Society Forum, Canberra. <www.natsoc.org.au>

Collignon, P. (2002). Antibiotics, livestock feed and human health. In *Good Grub – Food for Healthy People and a Healthy Planet.* (Ed. B. Furnass). pp. 93–95. Nature and Society Forum, Canberra. <www.natsoc.org.au>

Krockenberger, A.K., Kitching, R.L. & Turton, S.M. (2003). *Environmental Crisis: Climate Change and Terrestrial Biodiversity in Queensland.* Cooperative Research Centre for Tropical Rainforest Ecology and Management. Rainforest CRC, Cairns (30pp).

Meadows, D.H., Meadows, D.L., Randers, J. & Behrens, W.W. (1972). *The Limits to Growth.* Universe Books, New York.

NSESD (National Strategy for Ecologically Sustainable Development). (1992). <http://www.deh.gov.au/esd/national/nsesd/>

Raskin, P., Banuri, T., Gallopín, G., Gutman, P., Hammond, A., Kates, R. & Swart, R. (2002). *Great Transition: The Promise and Lure of the Times Ahead.* A report of the Global Scenario Group. Stockholm Environment Institute, Boston.

Stanley, F. (2002). Child health since Federation. *Year Book of Australia 2002.* Australian Bureau of Statistics.

Thomas, C.D., Cameron, A., Green, R.E., Bakkenes, M., Beaumont, L.J., Collingham, Y.C., Barend, F., Erasmus, N., Ferreira de Siqueira, M., Grainger, A., Hannah, L., Hughes, L., Huntley, B., van Jaarsveld, A.S., Midgley, G.F., Miles, L., Ortega-Huerta, M.A., Peterson, A.T., Phillips, O.L. & Williams, S.E. (2004). Extinction risk from climate change. *Nature* **427:** 145–8.

Wackernagel, M., Schulz, N.B., Deumling, D., Linares, A.C., Jenkins, M., Kapos, V., Monfreda, C., Loh, J., Myers, N., Norgaard, R. & Randers, J. (2002). Tracking the ecological overshoot of the human economy. *Proceedings of the National Academy of Sciences* **99 (14):** 9266–71.

Wallstrom, M., Bolin, B., Crutzen P. & Steffen, W. (2004). The Earth's life support system is in peril. *Herald Tribune.* 20 January 2004.

Western Australian Government. (2003). *Hope for the Future: A Vision for Quality Life in Western Australia.* (Prime authors: Peter Newman and Michael Rowe) Department of the Premier and Cabinet, Perth.

2 – Sustainability, health and well-being

Tony McMichael

We organise societies, run economies and build infrastructure to make life better, healthier and longer for humans. What is widely missing from some of our mainstream disciplines, including economics and demography, is the recognition that human sustenance, environmental stability and the flow of materials and other 'services' from nature are absolutely central to good health, survival and social advance. The Earth's ecosystems are the life-support systems for all human societies and the basis of our economic life, and we disrupt them at our peril. Recently, the idea has emerged that health is largely a matter of individual choice, behaviour and access to health care. But it is population level shifts in human culture, technology and environmental demands that have, throughout history, altered the patterns of well-being, disease and survival. We must now conduct our collective living on the non-negotiable terms of the natural world.

The debate about population size, environmental management and human well-being in Australia can no longer ignore the fundamentals of 'sustainability'. There is now worldwide (including Australian) evidence that the human enterprise is increasingly overloading the capacity of the natural world to supply, replenish and absorb – and recognition of this necessarily brings a new dimension to this public discussion (Vitousek 1997). Henceforth, that discussion about the sustainable management of population and environment must refer to the indefinite, or at least very long-term, human carrying capacity of the Australian environment, and of the biosphere at large. We must recognise that we are now seriously, though largely inadvertently, disrupting aspects of the planet's life-support system. This has obvious implications for the sustainability of good health in current and future human populations.

Since human experience – that is, happiness, fulfilment, well-being and health – is actually the bottom line of 'sustainability', it is important to understand the ways in which impairment of both environmental and social sustainability affects human experience (McMichael 2002). In this chapter I concentrate on how non-sustainable structures and practices will affect risks to health, and consider the basic strategies we must develop to lessen those risks to population health.

This perspective is not yet widely appreciated. Much of the current discourse about sustainability refers to more obvious, readily tangible, entities – economic indicators, infrastructural assets, recreational facilities, stocks of species, and so on. But, from a human-centred viewpoint, these are not *ends*; they are *means* to human experiential ends. We actually organise societies, run economies and build infrastructure to make life better, healthier and longer for humans.

Such an understanding will not come easily. Most of us (particularly the older generations of Australians) are hampered by conceptual blinkers. The education system that most of us passed through during the latter half of the 20th century was systemically blind to the fundamental dependence of human societies on their natural resource base. The dominant scientific disciplines at school and university were those that had underpinned the industrial revolution – physics, chemistry and mathematics. Biology was also taught, but as an empirical science with little reference to the principles of evolution or ecology. It is little wonder that, today, we have scant appreciation of the interconnectedness, the ecological processes and the limits of the biosphere.

This non-receptiveness to ideas of ecological interconnectedness, of limits to 'growth', and of sustainability in general, characterises many of the key research disciplines that bear on social planning, environmental management and policy making. Consider that neither mainstream economics nor demography incorporate any real appreciation of environmental constraints into their thinking. Rather, both disciplines view the world as if it were an open system, within which their discipline-specific processes occur freely. Although economists have broadened their perspectives in recent years, many still use a conceptual framework that excludes consideration of both human influence and human dependence on ecosystems. Contemporary epidemiologists primarily focus on individual-level behaviours and circumstances as causes of disease, thereby overlooking the broader ecological dimension which draws attention to the underlying social, cultural and political determinants of the patterns of disease risk within and between populations. Political science is preoccupied with issues of national security, social stability, the balance of tensions between neighbouring countries in relation to resource access and population movement, and some issues of international ethics and moral responsibilities. However, it does not normally address questions about forms of social organisation appropriate to the achievement of sustainability.

What, then, is widely missing from these various disciplinary perspectives is recognition that human sustenance, environmental stability, and the flow of materials and other 'services' from nature are the foundation of good health, survival and social advance. Earth's ecosystems are the life-support systems for all human societies and are the basis of our economic life. We weaken or disrupt them at our long-term, if not immediate, peril.

Views of 'human health' vis-à-vis 'sustainability'

On the above argument, the improvement and maintenance of human population health should be a central criterion of 'sustainability' (McMichael 2002). Changes in population health outcomes over time, interpreted appropriately, will convey more useful and relevant information about the sustainability of society's chosen path than can the conventional, but often seriously misleading, indices of economic performance and wealth levels.

Genuine sustainability, which spans distant future generations, requires that we achieve societies able to maintain both the natural resource base and internal social cohesion. However, achieving those conditions will require unprecedented enlightened stewardship of the biosphere. There is mounting evidence that humankind, via its expansions in numbers and economic intensity, is now overloading the biosphere (McMichael 2001b).

The world's climatologists have recently converged on the view that we are now experiencing the effects of human-induced climate change (IPCC 2001; see also Chapter 7, this volume). This is just one symptom of planetary overload. The others include the damaging of stratospheric ozone (with resultant increased ultraviolet radiation flux), the degradation of productive land, the disturbance of some of the world's great elemental cycles (sulphur, nitrogen, phosphorus), the depletion of freshwater supplies, the weakening of many of the ocean's great fisheries, and the global dispersal of various persistent (mostly chlorinated) organic chemicals (Raven 2002; McMichael 2001b).

Meanwhile, in modern Western society in which the combination of individualism, consumerism and market economics is the predominant ethic, we have also a commodity-like view of 'health'. That is, we see our health as being largely transacted at a personal level via individual choice, behaviour, and access to health care. Yet it is actually the population-level shifts in human culture, technology and environmental impacts that, over the centuries, have repeatedly altered the patterns of well-being, disease and survival. That is, the levels and types of disease in a population are predominantly determined by larger-scale ecological and social influences.

Epidemiologists have been caught up in this same shift of focal length, moving from group-level to individual-level emphasis. The original discipline, in mid-19th century Europe, focused on describing and explaining health differentials between geographic regions and population subgroups. Subsequently, as infectious diseases were replaced by non-communicable diseases (mostly of later adulthood) in Western populations during the latter years of the 20th century, epidemiologists have refocused their research attention on elucidating individual behaviours and experiences as the 'causes' of disease. Populations have thus come to be viewed primarily as unstructured aggregations of individuals exercising free choices as consumers and citizens.

Recently, there have been corrective influences to this narrowed orientation of epidemiological research. In particular, there has been a restitution of interest, led by the 'social epidemiology' subdiscipline, in studying the more 'upstream' influences on disease causation: that is the social, economic and cultural factors that determine the actual distribution of disease within populations, and the overall population rate of disease (Eckersley *et al.* 2001).

The resurgence of infectious diseases has re-emphasised population-level phenomena, including cultural practices and technological choices. The dramatic downturns in health and life expectancy in the ex-Soviet bloc countries, following the collapse of communism, highlighted the fundamental importance of social, economic and political conditions and institutions. The rise of obesity in modern urban populations everywhere provides a ready example. We cannot sensibly tackle it at the level of individual genes, eating behaviours or personal activity levels since it is a manifestation of radical changes in average ways of living. In Australia, as in other modern affluent societies, the availability of processed energy-dense foods has increased over recent decades, as has the tempo of commercial advertisement (especially directed to children and young people). To compound the effect, with labour-saving devices at home, at work and to transport us, and with television and electronic gadgetry to distract us from outdoor recreation, our daily level of physical activity has steadily fallen. The net result is a systemic daily energy imbalance leading inexorably to a rising prevalence of overweight and obesity. Fifteen years ago, around one-quarter of adult Australians were classified as overweight; today the figure is around half, and rising. Accordingly, we need to change societal priorities in food production and consumption, and to design cities and workplaces in ways that ensure that adequate physical activity is undertaken. Meanwhile, there is nascent recognition of the risks to human health – both now and in the future – posed by global climate change and other global environmental changes, and associated developments in formal research methods (Martens *et al.* 2002).

So how should we view health in relation to sustainability? We may think of population health as an input to economic development (McMichael 2002). This, while partially relevant, is a utilitarian view of health. We may express concern over the health risks, physical and mental, attendant upon industrialisation and economic growth. This is a janitorial approach, seeking to tidy up after society. Or, unusually, we may think in larger-framed, ecological, terms, and treat population health as a centrepiece of the Sustainability Debate.

Viewed within this frame, we can appreciate that some of the celebrated health gains of recent decades have been achieved partly at the expense of the world's natural capital. This is not easy arithmetic to do, nor are all the costs incurred directly. The following examples illustrate some of the likely 'downside' aspects (mostly ecological accounts that are yet unpaid) of recent gains in life expectancy.

- The agricultural 'Green Revolution' that occurred in many developing countries during the 1970s and 1980s via the planting of high-yield grain crops and increased use of irrigation, fertilisers and pesticides enhanced local food security. However, in the process it damaged much arable land via irrigation, fertilisation, compacting and erosion.

- Harvesting wild fish for a burgeoning world population has depleted various ocean fisheries. We will need, increasingly, to rely on fish-farming to maintain supplies.
- Injudicious antibiotic use, including for the production and maintenance of domesticated animals and houseplants, has rendered many pathogens antibiotic-resistant.
- Food preservation with chlorofluorocarbon-based refrigeration has contributed to stratospheric ozone destruction.
- The powering of cities (with its undoubted health benefits) has contributed to greenhouse gas emissions.

Overall, then, appreciation of the basic ecological significance of population well-being and health, viewed within a larger frame and over decadal time, will enrich our understanding of both the goals and guidelines for 'sustainability'.

Let us consider in a little more detail, and within a broadly based ecological framework, two of the above-mentioned issues: emergent infectious disease, and the health risks posed by global environmental changes, especially climate change.

Infectious disease ecology

There has been a recent upsurge of interest in the apparent rise, or increased adaptability, of infectious diseases. It is no surprise, for example, that our widespread overuse and misuse of modern antibiotics has led to a general increase in microbial antibiotic resistance. We have been putting intense selection pressure on microbes, with these antibiotics, and they – as they have been doing for several billion years in nature – have responded by acquiring metabolic resistance to these otherwise lethal chemicals. With a generation time of mere hours, it is inevitable that, among the millions of mutant microbes produced daily, there will be some that, by random good fortune (for the microbial species) are resistant to any particular antibiotic.

Infection is intrinsic to life on Earth. The bacteria, viruses, protozoa and other infectious agents obtain nutrients and energy by parasitising higher organisms. Most infection is benign, some is beneficial to both host and microbe, and some adversely affects the host's biology – it is these that we call 'infectious disease'. During the long processes of human cultural evolution, population dispersal around the world, and subsequent inter-population contact and conflict, there have been several distinct transitions in the relationships of *Homo sapiens* with the natural world (McMichael 2001a). Each of these transitions in human ecology and in inter-population interaction has profoundly changed the patterns of infectious disease.

The three great historical transitions since the early emergence of agriculture and livestock herding from around 10 000 years ago occurred when:

- i early human agrarian-based settlements enabled various species of animal-infecting microbes (usually as mutant strains) to enter *Homo sapiens*, thereby creating new human diseases such as smallpox, measles, typhoid, tuberculosis, leprosy, the common cold, influenza, and dozens of others;
- ii early Eurasian civilisations (such as the Roman Empire, China and South Asia) came into military and commercial contact, around 2000 years ago, swapping their dominant infections; and
- iii European expansionism, over the past five centuries, caused the transoceanic spread of oft-lethal infectious diseases.

This latter transition is best known in relation to the conquest of the Americas by Spanish conquistadors, when the inadvertent spread of measles, smallpox and influenza decimated the vast native Amerindian populations of Central and South America. Similarly, the introduction of smallpox (and other infectious diseases) to Australian Aboriginals soon after European settlement caused local devastation of many Indigenous communities.

Today, we may be living through a fourth great transitional period. The contemporary spread and increased adaptability of various infectious diseases, new and old, reflects the impacts of demographic, environmental, behavioural, technological and other rapid changes in human ecology. Within the realm of clinical medicine, there are various manoeuvres that we now routinely use that have provided very inviting opportunities for opportunistic microbes. These include blood transfusion, organ transplantation and the widespread use of hypodermic syringes. These have contributed to the rising iatrogenic (caused by medical examination or treatment) problems of hepatitis C, HIV/AIDS, and several other viral infections. The increasing problem transmission of hepatitis C via blood transfusion in Australia, identified during the 1990s, has necessitated new procedures for screening blood donation.

As with previous human–microbe transitions, a new state of equilibrium may lie ahead. However, it certainly will not entail a world that is free of infectious diseases. Any future sustainable human ecology will have to come to terms with the need for, and hence the needs of, the microbial species that help to make up the interdependent system of Life on Earth.

Let us explore this ecological dimension further, especially in relation to our growing impact on the world's environment. This dimension refers to the larger-scale population-level changes, mentioned above, that exert the important long-term influences on the patterns of disease and survival in populations.

Changes in human ecology, environmental impacts and risks of disease

As we have seen above, a succession of profound changes in human ecology has occurred over the past one hundred centuries, from the time that some of our hunter-gatherer forebears began to seek out more secure sources of food via the early processes of farming and herding. Debate continues about the reasons for that radical change in the primitive human economy, but its immediate consequence was indisputable: the local carrying capacity of the environment was increased via this human intervention, and agrarian populations duly expanded. The main changes in human ecology over the ensuing 10 000 years have been particularly in food production, social structures, urban living, industrialisation, reproductive behaviour, and, now, demographic profile (as populations undergo an unprecedented ageing).

The career of *Homo sapiens*, viewed at the global scale, has now reached a critical juncture in another very important respect (Raven 2002). Population numbers have quadrupled over the past century; the scale of economic activity has increased approximately twentyfold over the same period; and, consequently, we have inadvertently begun to change the conditions of life on earth by altering the global climate, depleting stratospheric ozone, extinguishing whole species and their local populations, and damaging the food-producing ecosystems on land and at sea. These global environmental changes, unprecedented in human experience, are an unavoidable result of, jointly, these increases in the size of the human population and in the intensity and type of economic activity.

There has been an increasing awareness of the growth in size of humankind's 'ecological footprint' in recent decades. The metaphor refers to the fact that a typical modern community depends on a very large area of Earth's surface to provide its material and energy needs and to absorb its wastes. For large cities such as London this 'footprint', or ecological overshoot factor, is around 100–200 times the size of the city itself. For example, at a national level, it is estimated that the Dutch population requires an area of Earth's surface approximately 15–20 times greater than the area of the Netherlands. At the personal level, each Australian currently requires, on average, a land-surface area of eight hectares to supply their food, energy and other materials and to absorb their wastes. Unlike the Netherlands, our population size is – at the moment – approximately in balance with the amount of biologically productive land surface in Australia (although we are clearly mismanaging freshwater supplies and losing species of animals and plants). Globally, however, if resources were shared both equally and sustainably, the available area per person for a population of 6.4 billion would be around two hectares. That is, we in Australia currently use much more than our fair share of Earth's resources.

Some very recent research has attempted to quantify the time-trend over the past four decades in humankind's ecological overshoot. Globally, it looks as if we have been operating 'in the red' for the past 20 years or so. Currently we are extracting from the biosphere, annually, a total of natural 'goods and services' that is about 25–30 per cent higher than can be provided sustainably (Wackernagel *et al.* 2002). In other words, we are now into the realm of ecological deficit budgeting, and, with current technologies and social priorities, we can only balance the books annually by continuing to deplete natural capital. This necessarily means that we are transmitting to future generations a partially depleted and less life-supporting biosphere. This has wide-ranging consequences for the future prospects of population health.

Global climate change and health

The best-known example of these modern global environmental changes is climate change. There has been a clear trend in global warming over the past three decades, and climatologists are now confident that most of this increase has been due to human influence on the composition of the lower atmosphere. Further, it has become apparent that, in recent decades, many non-human physical and biological systems have undergone alterations that are reasonably attributable to climate change.

The human species, via social organisation and cultural practice, is much better buffered against environmental stressors than are all other plant and animal species. Hence we should expect *Homo sapiens* to be a less sensitive early-responder species. Further, there is no simple description of the current and potential future health impacts of climate change. Not only will climate change induce different types of environmental and ecological changes in different geographical regions, but also the vulnerability of human populations varies – as a function of locality, level of material resources, technological assets and type of governance.

There is a wide range of expected health impacts of climate change (McMichael *et al.* 2003). These do not entail novel processes and unfamiliar health outcomes. Rather, they entail climate-induced changes in the frequency or severity of familiar health risks – such as floods, storms and fires; the mortality toll of heatwaves; the range and seasonality of infectious diseases; the productivity of local agro-ecosystems; the health consequences of altered freshwater supplies; and the many repercussions of economic dislocation and population displacement. Most of the expected health impacts will be adverse; a few will be beneficial.

There is already some suggestive evidence that climate change is affecting particular health outcomes. Some vector-borne infectious diseases – such as tick-borne encephalitis in Scandinavia, and malaria and perhaps dengue fever in tropical and subtropical regions – have increased their geographic range or seasonality in ways that accord with the observed changes in climatic conditions over the past two decades. Cereal grain yields have become a little more unstable during the late 1990s and into this decade, initially displaying increased inter-annual variability and having declined monotonically over the past six to seven years. Could this be, at least partly, due to changing climatic conditions? Extreme weather events appear to have increased in tempo during the 1990s, with predictable impacts on human life and limb. Several small island states are experiencing growing concern about sea-level rise. Even at this early stage, such concern may be jeopardising inhabitants' well-being and mental health.

A paradox?

There seems to be a paradox in this current human–environment relationship. If environmental conditions and ecological systems are important to human well-being and health, how can the world's environmental indicators be declining while life expectancy is rising? Further, is it true that the unplanned, rapid growth of most of the world's cities entails a loss of social cohesion and a deficiency in 'social capital' – that is, a loss of institutions and community structures that help society and its component communities to work harmoniously and productively? If so, why is that not also affecting population health and well-being?

There are three possible explanations for this apparent paradox:

i modern human societies, via technological, economic and political achievements, have acquired near-immunity to adverse external environmental circumstances;
ii adverse health effects *are* already occurring, but we have not yet detected them. Not only is the real world of disease causation multi-causal and 'noisy', but we have no null comparison data (that is no global population living in an unstressed environment);
iii there is a lag period between the decline in environmental conditions and the occurrence of health impacts. This lag reflects both complexity of process and the protective buffering afforded by human culture.

The first explanation is not tenable – and, anyway, it is countered by long human experience. Throughout history, even as technologies have advanced, great civilisations, as in Mesopotamia, Egypt (the Old Kingdom), the Indus Valley, Mesoamerica, Peru and elsewhere, have crumbled in the wake of environmental infrastructural decline (McMichael 2001b). Today, various countries, including China and Australia, are experiencing the adverse environmental consequences of land clearing, over-exploitation of arable land and excessive reliance on irrigation. Underlying all such accounts is the fact that human societies are necessarily beholden to ecological limits and the laws of thermodynamics, and therefore cannot live apart from the natural world. That is, the human economy is a wholly dependent subset of the natural 'economy'. Therefore, in the longer term, if not sooner, there can be no immunity of modern human societies to the consequences of environmental degradation and the weakening of ecosystems. Most probably, then, the best explanation for the apparent paradox is a combination of the second and third items above.

Future prospects for population health

Over the next few decades, life expectancies will probably continue their historically unprecedented rise, especially in low-income countries as they complete the process of health transition. However, if the HIV/AIDS pandemic intensifies, then life expectancies will decline in afflicted nations as they have already done in many sub-Saharan African countries. Some commentators point out that the recent emergence of HIV/AIDS may presage the emergence of other new infectious diseases, as patterns of human living, mobility and behaviour continue to change rapidly. We cannot know these things in advance, but we should expect the microbial world to take advantage of new opportunities that human cultural evolution provides.

On current trends, globally, the proportion of deaths from infectious diseases is projected to halve from around one-third to one-sixth of total deaths, whereas the proportion due to coronary heart diseases, stroke, cancer and other non-communicable adult diseases will rise from around one-half to three-quarters (WHO 2003a). The proportion of deaths from injuries, too, will increase. Malnutrition and unsafe drinking water in the less-developed countries, along with indoor air pollution from cooking and heating in poor households, will remain major killers – even as cigarette smoking, alcohol consumption and dietary excesses cause increasing rates of adult disease and premature death.

The burgeoning global tobacco epidemic killed at least 4 million people in 2000, and by 2020 it will be killing approximately 10 million people per year – that is, about one in every three adult deaths. Diabetes type 2 (the most prevalent type, with long-term adverse health consequences) currently afflicts around 4 per cent of the world's adults, and is becoming more prevalent as urban populations everywhere get older and fatter. The current 120 million cases worldwide will approximately double over the coming decade. Australia, with its relatively high prevalence of obesity and diabetes type 2 (AIHW 2002), faces a similar prospect where our approximately half-million cases today will become a million cases by around 2015.

The widespread decline in traditional family and social supports, amplified by the urbanisation and increasing mobility of modern populations, may contribute to mental depression becoming a major source of chronic health impairment within several decades. The World Health Organization has projected that mental depression will have become one of the several major causes of poor health and disability by the year 2020 (WHO 2003a).

In this changeable modern world, surprising shifts in disease patterns may become more frequent. In most countries, life expectancy has continued to rise over the past several decades. In Australia in 2001, life expectancy at birth reached 77 years for males and 83 years for females. But not all countries have experienced gains. Life expectancy plummeted in Russia in the early 1990s as social structures and controls dissolved following the collapse of communism. Elsewhere during the 1990s, adult life expectancy fell by at least two years in around 10 other (non-African) countries, including Haiti, Ukraine, Moldova, North Korea and several countries of Central Asia (McMichael *et al.* 2004).

Infectious diseases are almost certain to spring some future surprises, despite views expressed prematurely around three decades ago by various eminent scientists to the effect that we had brought infectious diseases under control with vaccination, antibiotics and better environmental management. The newly named variant Creutzfeldt-Jakob disease (human 'mad cow disease') appeared unexpectedly in Britain in the mid-1990s, and its future course remains uncertain. HIV/AIDS emerged during the 1980s and, by the year 2000, was killing over two million people annually. Cholera has extended its dominion over the past quarter-century, having embarked on its longest-ever pandemic. Tuberculosis, assisted by HIV, has rebounded within poor and malnourished populations around the world. During that same period, the mosquito-borne diseases, malaria and dengue fever, have been resurgent. Dengue fever, substantially under control in the 1970s in Latin America, has subsequently re-established itself widely in that region, infecting more people annually than ever before.

Ways of getting there

The task we face in achieving an ecologically and socially 'sustainable' society is unfamiliar, unprecedented and formidable. But our latent brainpower, to be applied to the task, is also formidable. There are three areas in which we need to change our current practice.

i *Research*: Recognition of the complex environmental and social problems that we face requires a widened interdisciplinary research capacity, including the application of complex systems science to these issues. This added-value model of science is well illustrated in the ongoing work of the UN's Intergovernmental Panel on Climate Change (IPCC). In 2004 the Australian Academy of Science convened a conference specifically to consider how best to achieve cross-disciplinary engagement in relation to big and complex issues to do with population, environment and human well-being. It is clear that the existing, traditional structures and tertiary-educational training constrain effective inter-disciplinary approaches. Nascent moves in Australian universities to create more interdisciplinary courses and research endeavours need to be reinforced – and supported by funding incentives.

ii *Education*: The oncoming generation of children must better understand the ecological framework within which the human species lives. The essence of 'sustainability' is that we must learn to live on the natural world's terms, not on our own presumptuously detached (and ultimately destructive) terms. That will require some profound change in the science curriculum of school education in Australia, as elsewhere.

iii *Politics*: We must move beyond our current 'adolescent' phase of human behaviour (Cocks 2003). Today's competitive, narrowly self-interested, nation-states are a modern analogue of ancestral warring tribes of hunter-gatherers. Awkwardly, this competitiveness, selfishness and 'short-termism' is deeply programmed into the human species, as the product of the long evolutionary struggle for survival. We must, therefore, deploy our self-awareness and our capacity for cultural sophistication, particularly our unique (though largely latent) ability to anticipate and shape the distant future, in order to override these more primitive, shortsighted, living-in-the-moment attitudes.

Increasingly, enlightened political leadership will attract votes from those who want a safer world for their grandchildren. But this process needs to start happening this decade. For the moment, Australian policy-making is largely in thrall to the immediacies of economic growth, consumer satisfactions, and the external threats of 'terror' – the latter tending to have us follow reflexively, the lead of our strong ally, the United States. Policy-making in relation to longer-term sustainability issues has largely been sidelined, and there has been little readiness to make an overt commitment to international efforts such as the Kyoto Protocol for reducing greenhouse gas reductions.

Underlying these three strategies must be a new awareness that 'sustainability' is ultimately about optimising human social and biological experiences. That may require changes in social and political organisation, and in the way we design and manage our communities (especially our modern large urban environments). It will certainly require that we conduct our collective living on the non-negotiable terms of the natural world. We can modify and reshape environments, we can develop clever technologies that extend the carrying capacity of the natural world, but the bottom line is that we cannot exceed the biocapacity of the planet without long-term detriment to human health and survival.

References

AIHW (Australian Institute of Health and Welfare). (2002). *Australia's Health 2002.* AIHW, Canberra.

Cocks, D. (2003). *Deep Futures.* UNSW Press, Sydney.

Eckersley, R., Dixon, J. & Douglas, B. (eds). (2001). *The Social Origins of Health and Well-being.* Cambridge University Press, Cambridge.

Intergovernmental Panel on Climate Change (IPCC). (2001). *Climate Change 2000.* Cambridge University Press, Cambridge.

Martens, W.J.M. & McMichael, A.J. (eds). (2002). *Climate, Environmental Change and Health.* Cambridge University Press, Cambridge.

McMichael, A.J. (2001a). Human culture, ecological change and infectious disease: are we experiencing history's fourth great transition? *Ecosystem Health* **7:** 107–15.

McMichael, A.J. (2001b). *Human Frontiers, Environments and Disease: Past Patterns, Uncertain Future.* Cambridge University Press, Cambridge.

McMichael, A,J. (2002). The biosphere, human health and sustainability. *Science* **297:** 1063.

McMichael, A.J, Campbell-Lendrum, D., Corvalan, C. *et al.* (eds) (2003). *Climate Change and Human Health: Risks and Responses.* World Health Organization, Geneva.

McMichael, A.J., McKee, M., Shkolnikov, V. & Valkonen, T. (2004). Mortality trends and setbacks: global convergence or divergence. *Lancet* **363:** 1155–59.

Raven, P.H. (2002). Science, sustainability and the human prospect. *Science* **297:** 954–6.

Vitousek, P., Mooney, H.A., Lubchenco, J. & Melillo, J.M. (1997). Human domination of the Earth's ecosystems. *Science* **277:** 494–99.

Wackernagel, M., Schulz, N.B., Deumling, D., Linares, A.C., Jenkins, M., Kapos, V., Monfreda, C., Loh, J., Myers, N., Norgaard, R. & Randers, J. (2002). Tracking the ecological overshoot of the human economy. *Proceedings of the National Academy of Sciences* **99(14):** 9266-71.

World Health Organization (WHO). (2003a). *Global Burden of Disease 2000.* World Health Organization, Geneva.

World Health Organization (WHO). (2003b). *World Health Report 2002.* World Health Organization, Geneva.

3 – Inequality and conflict

Colin Butler

Darwin's concept of fitness implies inequalities and competition as the main drivers of the evolutionary process. Yet cooperation within and between species is essential to achieve stable societies, a process which is supported by complexity theory. Over recent years there have been increasing inequalities in access to economic and natural resources between and within industrialised and developing countries. Recognition of these disparities by the dispossessed, combined with resource depletion and population growth, promote ethnic and religious discord and violence that can only exacerbate the situation. Achievement of peace and sustainability for our precarious world will require a revolutionary change away from denial and towards widespread consideration of carrying capacity and inequality, rather than aggravating instability through policies embodied in the so-called 'War on Terror'. This will entail an ideological shift from competition towards cooperation.

Inequality characterises many human and non-human systems. It can never be eliminated, but this does not necessarily mean inequality is evil, immoral or wrong. Some inequality – probably a substantial degree – is required to provide incentives, make rewards meaningful and to optimise production, including of many public goods. Having less than others is not necessarily wrong in itself, even morally, particularly if the wider system allows prospects of individual improvement and reward for effort. However, beyond thresholds, excessive inequality is a moral, economic and political problem. Imagine a banquet in which a small number of gluttons are waited upon by a large number of enslaved but half-starved attendants. Outside, equally starved troops guard the hall. Most people would react to this fantastic situation not only with revulsion but also with incredulity; a moment's thought would suggest the situation is unstable, if not through poisoning then through an organised revolt by the waiters and troops. It is likely that after this revolt another unequal society would form, but one which would be – even if only slightly – less unequal.

The extent of inequality in any system; including human society, is *always* limited. Like most phenomena, the degree of inequality fluctuates and at times will become sub-optimal – either too equal or too unequal. Nonetheless, mechanisms exist to constrain this oscillation within limits. Within human societies, these checks and balances include reputation, rebellion and mutual gains from cooperation (Wright 2000).

With hindsight, the French Revolution has an air of inevitability due to the excesses of Mary Antoinette's court, which are still a part of folk memory after more than two centuries. On the other hand, the system of democracy and free enterprise that characterised the US during the 20th century undoubtedly contributed to that country's economic ascendancy, combined with its access to vast natural resources. At the same time, the stifling of freedom in the former Soviet Union constrained that economy's growth, in part because of reduced domestic inequality. Adam Smith, a father of modern economics, commented on how the lack of property rights in Turkey stifled that country's growth – there being so little protection for individuals who showed initiative and enterprise that enterprise went unrewarded. But Smith also warned against the equally stifling effects of monopolies, which if left untouched could function like a despotic ruler.

Today we live in an interconnected world. Many of the goods we use, much of the food we eat, and all of the air that we breathe has been produced, grown, or modified by people and economies from all over the planet. This matters for inequality. Long ago, the poor in our society lived in our proximity. A combination of compassion, shame and fear restricted the degree to which the wealthy could oppress or ignore the poor. Later, the emergence, in many countries, of a free press and of the vote further guaranteed that the poor could not fall too far behind the wealthy. But now, in our globalised world, most of the poor live out of sight, and usually out of mind. These people – of whom there are literally billions – may be peasants in sub-Saharan Africa, flood refugees in Bangladesh, or tribal people in Paraguay. We know little of them; they know little of us. Yet increasingly we are connected, not only economically, but through gradually tightening information chains.

Of course, these economic connections are indirect. The poorest people in Bangladesh are unlikely to be working in the textile factory that may have made your shirt, but the *comparatively* comfortable living standards of the workers in a Dhaka factory may depend, through only a few more steps, upon the poverty of that person. We are all part of a vast economic food chain. I, the writer, and you, the reader, live towards the top of this pyramid, though we are only dimly aware of the apex above us. Way below us there lives the textile worker – yet her fingers may have touched my shirt. And only a few rungs below her there live people who are truly in abject poverty, experiencing extremely restricted, vulnerable and powerless lives.

This economic pyramid has been growing in size and height for centuries. Today the magnitude of inequality in the global system is unprecedented, and it is breathtaking. Five hundred years ago, kings were still prone to diseases that are today easily curable, though they were probably always spared from hunger. Today, despite our knowledge of vaccines, antibiotics, education, and our vast *collective* wealth, hundreds of millions of people continue to miss out, in ways that would be regarded as totally unacceptable if they were either more visible, or more organised and threatening.

Of course, there are complex reasons for inequality, and many causes are far from tractable. But, for a few decades, especially after World War II, substantial progress was made in making the world fairer. A combination of enlightened self-interest, idealism, new technologies and Cold War competition saw vast improvements in the living conditions in many parts of the Third World. Life expectancies rose dramatically, roads were built, literacy increased and birth rates fell quickly. The Green Revolution assuaged the fears of famine that had risen to prominence in the 1960s.

Sometime around 1980 much of this progress petered out. The reasons for this include economic shifts in the West, as stagflation and an overly trusting Western population allowed the installation of policies that placed a greater emphasis upon individual than collective goods. Originally these policies were proclaimed as the new, smart, way to *reduce* inequality and to improve public goods, both nationally and globally. The wisdom of the market was alleged to automatically result in greater total wealth, which would result in a beneficial 'trickle down' effect, reaching not only our own poor, but also the poor in the Third World. Consequently, it was considered rational and humane to slash foreign aid and reduce subsidies that lowered the costs of Third World schooling and health care. We were assured this would accelerate development.

Two decades on, global inequality has substantially clearly increased (Krugman 2002) and foreign aid from Western societies remains low and, despite a few valiant rearguard actions, unfashionable. In some cases – especially in China and East Asia, where human fertility was already low and where a great emphasis continued to be placed upon education – economic takeoff was achieved. Although China, in particular, has vast environmental problems it stands a good chance of continued lift-off, even if climate change increasingly bites.

But in much of sub-Saharan Africa the application of these new policies has been truly catastrophic. Populations in a host of African countries have experienced massive declines in life expectancy, deepened food insecurity, appalling governance, and periodic and barbaric conflict. South Asia is also not out of the woods. Two countries rich enough to bear nuclear arms – India and Pakistan – also contain populations with substantial hunger, high maternal and child mortality rates, regional political instability and patchy governance. They also still have worryingly high birth rates and high inequality.

Brazil – a land of kidnappings, excess, gated communities and contemporary slavery (Bales 1999) – is often considered to house the most unequal single national economy. In parts of Sao Paulo, Brazil's largest city, a new caste is arising, where security guards employed in the gated communities need to demonstrate that their *parents* have a clean police record. Now, a Left-leaning government has been elected, and inequality is likely to fall, bettering Brazil's long-term prospects.

If the world were a single country, its inequality would make Brazil look fairly egalitarian. This is true no matter how inequality is measured. A currency known as international dollars which adjusts for purchasing power increases the income of the poor in developing countries. If inequality is measured using an internationally traded currency, such as US dollars, inequality is even more pronounced (Butler 2000). A vigorous debate rages in academic circles over the virtues and failings of these competing measures of inequality, yet neither side disputes the central fact that global inequality between countries exceeds that which exists inside any nation (Wade 2004).

Debate also rages over the extent to which inequality matters. Defenders of high inequality argue that a rising tide lifts all boats, and that as long as real wages and living conditions improve for the poor it is irrelevant if they fall even further behind. This argument reflects ignorance of human nature. Increasingly, it also reflects ignorance of the facts; in many ways the tide is now receding for the global poor, who number in the hundreds of millions. Nonetheless, rebellion by the poor against global oppression depends upon more than the degree of inequality; it also depends upon the poor having means and opportunity.

But what has this got to do with sustainability? A great deal.

Limits to inequality

Cooperation and competition pervade human society. Although recent economic theories have tended to emphasise the virtue of competition, society is unimaginable without cooperation, such as for hunting and baby-sitting (even by non-relatives). The extreme position: that humans will invariably try to maximise their own wealth with little regard for others has been described as 'homo economicus', and was illustrated in the film *Wall Street*, which made 'greed is good' a household phrase. This position has little basis. Human societies display far more reciprocity and cooperation than greed and selfishness.

Reciprocity glues families, societies, and even markets, where buyers exchange cash for goods; if buyers feel cheated they will turn elsewhere. Reciprocity is often indirect. This may occur between individuals, families, and states. Consider the story of the old professor, who, when asked why she attended so many funerals, explained '... otherwise, no one will come to mine'. In this case, her act of courtesy cannot be returned by her recipients; she is instead relying upon indirect reciprocity. In the salmon-rich (and slave-holding) cultures of what is now called British Colombia, Amerindian societies held competitive potlatches – spectacular displays of wealth and generosity. These were expected to be reciprocated at an unspecified time, perhaps by the next generation.

Reciprocity, both direct and indirect, also characterises international relations. The Australian commitment of troops to both Gulf Wars can be considered as partial repayment for American assistance in World War II, as well as a down payment for future protection. By its alliance with a great power, Australia feels safer, purchasing short-term security. However understandable, this reciprocity suggests that Australia has internalised neither the reality of globalisation nor the threat of unsustainability. The lifestyle of both the US and Australia is contributing to the erosion of sustainability as well as to increased resentment, including within our region.

Not everyone is a reciprocator. Free riders – individuals who seek to benefit from the work of others without proportional reciprocation – are also common, though even free riders must reciprocate within subcultures. Free riders are often humiliated and vilified, and this signal provides a useful mechanism to reinforce cooperative behaviour by the majority (Bowles & Gintis 2002).

Reciprocity, both direct and indirect, forms the basis of all systems of morality (Nowak & Sigmund 2000). Rewards and punishment for good and bad behaviour are not just integral to Judaeo-Christian faiths but also to Hinduism and Buddhism. Reciprocating (whether a favour for a favour or an eye for an eye) contributes to a reputation. In the smaller communities from which modern civilisation has evolved, reputation can be considered as a naturally selected favoured mechanism to limit inequality. Reputation is a form of public information encapsulating myriad acts, not only by individuals but also by families, tribes and nations. Reputation provides a means by which strangers can penalise and reward past behaviour. A good reputation is a kind of passport, smoothing transactions with strangers. On the other hand, a poor reputation is like travelling in a car with a flat tyre. People are suspicious, unhelpful and sometimes hostile.

Groups also acquire reputations, but sometimes these may be undeserved. The reputation of the 'other' often acts as a barrier, reducing both cooperation and contact. This is especially obvious between groups with different languages, religions, customs and socio-economic status. Adverse reputation easily merges with caricature, prejudice and racism, and this can help to generate and sustain a self-fulfilling 'lock-in' where both sides hold extreme positions about the other, often over generations. Prejudice is likely to reduce contact between individuals in the characterised groups, leading to more segregation, and even fewer opportunities to weaken that prejudice.

At the same time, the shared customs and practices that may be regarded by outsiders as justifying contempt and prejudice (such as beef eating may appear to some Hindus) can foster and nurture support and cooperation within subgroups. Shared dress and slang may express solidarity, and, sometimes, nascent opposition against majorities perceived as oppressive. In France the dominant culture recently banned the wearing to school of headscarves by Muslim schoolgirls. It remains to be seen whether this will lead to more or less integration of Muslims among French society. It could easily lead to greater lock-in and more mutual resentment, unless parallel efforts are made to increase opportunities for Muslims to genuinely participate within French society.

In a small village the reputation of an individual is probably fairly accurate, though the reputation of *groups* (such as untouchables in India) may reflect the distribution of power and influence within that society, rather than impartial judgement.

In the global village, however, reputation and public opinion are increasingly manipulated, whether by propaganda or public relations. This relatively new phenomenon threatens the utility of reputation as an indicator of merit of this ancient social mechanism. The deliberate and large-scale manipulation of information, including of reputation, also threatens sustainability (Beder 1998). For example, the Australian academic Sharon Beder has documented many cases where representatives of the oil industry have massaged public opinion in ways that have attempted to cast doubt on the reality and severity of climate change. Tactics used to distort public opinion concerning environmental hazards range from the simultaneous publication of newspaper articles critical of the environmental movement to frivolous lawsuits intended to intimidate citizens' groups trying to protect environmental amenity (Beder 1998). Such campaigns should not be surprising, they belong to a long tradition where the pursuit of short term profit has over-ridden scientific opinion, from asbestos to lead and tobacco.

Justice, rebellions, and the war on terror

One person's 'inequity' is often another's 'justice'. Convicts were once transported to Australia in a time when the Antipodes must have seemed to the poor almost as remote as Mars does today, for crimes as trivial as the theft of a sheep. This justice may have seemed unfair, or inequitable to the individuals and families affected but was acceptable to many including, crucially, those with the greatest political power. This plasticity of justice is relevant to both the war on terror and to sustainability.

Evidence for the generally suppressed longing by the poor for greater equality can be traced in European history through the durable legend of Robin Hood, loved by the poor for his redistributive efforts, and before that, to the Arthurian legend of a round table, where no knight had precedence. Magna Carta, in which King John was forced to accede some rights to his nobles, also illustrates an early step in reducing British inequality. In the long march towards greater equality in Britain, force was often applied (by either side), but at other times the use of force was averted, because rational calculation suggested an overwhelming victory.

Most commonly, force was averted by the poor because it was abundantly clear that the means to guarantee success were deficient. But at certain favourable times, physical expressions of the desire to seek a fairer society were able to extend far towards the base of the economic pyramid. At a few times the poor could ask, like Oliver Twist, for more without expecting to receive instant and brutal repression in exchange. Prior to the Black Death, for instance, serfs were so controlled that they were required to keep their hair short as a visible mark of servility. But after the Black Death, especially in Western Europe, the bargaining position of the peasants increased substantially as the resulting shortage of labour improved their real wages.

One of the most famous peasant revolts was led by Wat Tyler, who marched upon London in 1381, soon after the Black Death. Initially the court was willing to negotiate with Tyler's men, but when the balance of power shifted slightly Tyler was promptly murdered. Nevertheless, within only a few more generations, serfs throughout Western Europe were liberated. By the time of the English revolution, in the mid-17th century, a group called 'the Levellers' were openly campaigning for ideals now recognisable as socialism. In the 19th century, it was the turn of the Chartists, who demanded an enlargement of the vote for men. Then came the suffragettes; gradually the franchise was widened to all adults in Britain.

It is not that the Black Death, the English revolution or the relative prosperity of the working class in Victorian times led to a greater desire by the poor for more equality, but instead that a slight relaxation in oppression allowed greater freedom to *express* this desire. In the lands that came to be known as the Third World soon after World War II, the period of decolonisation also permitted more open demands for freedom.

Global terrorism and its connections to unsustainability

The events of September 11, together with a host of less spectacular terrorist attacks, have come to define the new millennium. Writing in the *New York Times*, Thomas Friedman (Friedman 2004) claimed that the threat to civilisation represented by terrorists constitutes the third great totalitarian challenge to open societies in the last 100 years. Might it be possible that the massive increase in global inequality that has occurred in recent decades – together with increased means, opportunity and understanding by the poor of the immensity of inequality – has fomented the terrorism against which wealthy and powerful governments have declared war (Rogers 2000)?

Of course, very few representatives of privileged populations accept this analysis, including the Australian Government. For example, Friedman asserted that the attack on the US in September 2001 was purely one of Islamic totalitarians trying to impose their will. But like most Western analysts of the war on terror, Friedman evinces insufficient regard for the possibility that deeper causes may underlie the concerted attack on relatively powerful and privileged populations, now well under way.

Naturally, terrorists are denounced by the comparatively privileged populations that they threaten. Such unilateral views are typical of winners, who in most cultures inherit histories written to support, and belief systems designed to justify, the status quo. The US itself emerged as a separate nation only after rising against Britain, the great power of the 18th century. Within the 13 colonies that created the United States this revolution was widely perceived as just. However, a different view was certainly held in London.

Now in the 21st century, the US and its allies, including Australia, preside over and materially benefit from an economic and cultural empire that in some ways is comparable to that of the British in the 18th century, including a sense of manifest destiny and cultural superiority. For example, this empire, to which the privileged classes in China and India are now being admitted, consumes the vast majority of fossil fuel and is thus mostly responsible for the enhanced greenhouse effect. While some parts of the empire, especially in Europe, are concerned about climate change, official policy in both the US and Australia appears contemptuous, not only of the danger of climate change, but also of the case for populations in developing countries to emit more carbon, in order to develop. This contempt for developing countries is evident in the refusal of Australia and the US to ratify the Kyoto Protocol, arguing that it lets developing countries evade their responsibilities.

This empire also has an almost complete monopoly on the possession of weapons of mass destruction. In his 2004 State of the Union Address, US President George W. Bush accurately described nuclear weapons as instruments of mass murder. A careful listener would not imagine, from either the tone or context of his words, that the US possesses thousands of these weapons.

The arrogance and hypocrisy of this empire, regarding global inequality, climate change and who should possess weapons of mass destruction is scarcely visible within the empire, yet they are as obvious and inflammatory to the people characterised as terrorists and barbarians outside it as were the excesses of Marie Antoinette's court to the French peasants.

A more thoughtful path to understand and to investigate the root causes of terrorism seemed briefly possible. At the World Economic Forum, held in sympathy in New York soon after 11 September 2001, a wide array of commentators, from Bill Clinton to the managing director of the International Monetary Fund, argued that global inequality was a key factor in the emergence of global terrorism. Financier George Soros said '... the asymmetric threat from people who find the world in which they live in to be unacceptable means that terrorism and anti-globalisation violence can become manifestations of this frustration'. Bill Gates said '... people who feel the world is tilted against them will spawn the kind of hatred that is very dangerous for all of us'.

Today, only occasional rays of light penetrate this draped window. Conservative commentators and governments generally ignore the link between terrorism and inequality, or, if pressed, dispute it, pointing out, for example that Osama Bin Laden is very wealthy, and that most of the hijackers on September 11 were comfortably off. This interpretation reveals both a lack of historical understanding of previous revolutions, and little insight into how complex systems operate (Waldrop 1992).

Complex systems, inequality and emergence

Complex systems theory is a recently developed field of study that examines the interactions between the myriad elements of many systems, in ways which lead to self-organisation. In other words, complexity theory suggests that fairly simple rules of thumb can sometimes lead to phenomena that would appear, to the naïve observer, as unbelievably complicated and implausible. A classic example is a bee or ant's nest. Obviously, no single insect, including the queen, possesses the neural capacity to plan, construct and maintain a hive, yet the collective behaviour of millions of individuals leads, effortlessly, to a functioning, self-organised system. Some have suggested that life itself may be a self-organising phenomenon, inevitable when sufficient precursor molecules exist. Lovelock has famously suggested that the biosphere may also self-regulate (Lovelock 1988).

Complexity theory has, as yet, been little explored by social theorists, although recent work has analysed phenomena such as standing ovations, Mexican waves, job finding, and some fads and fashions as resulting from 'information cascades' which lead to either the fading or emergence of altered social states (such as the popularity of a hula hoop, whether to take a holiday in Tasmania and the offer of a job) (Miller & Page 2004). This theory gives insight into phenomena such as formal and informal social norms, but has not yet been applied to the emergence of revolts and revolutions.

Despite the lack of a supporting scientific literature, it is likely that many other social phenomena, including revolts, religious cults and genocides also illustrate key elements of complexity theory. It is plausible that revolts are an emergent, balancing mechanism operating to reduce inequality whenever it becomes excessive, triggered by motivation, opportunity, and a reasonable shot of success. Charismatic leaders may help initiate revolts, but self-organisation theory downplays the importance of individuals. Instead it argues that in favourable conditions many moderately well-suited individuals may catalyse effects made likely by causal factors of far more weight than any one person.

The attacks on the Pentagon and the Twin Towers in September 2001 can thus be viewed as an emergent phenomenon, arising through information cascades, and occurring in response to the reality and – importantly – the *perception* of global inequality. Removing individual terrorist cells in response is therefore unlikely to be successful, because it does nothing to solve the root causes. The current 'War on Terror' may even generate more terrorism over the medium term, including by diverting resources away from the Millennium Development Goals and strategies to ameliorate global climate change. These goals, announced soon before 11 September 2001, relate to quantitative targets to improve many aspects of Third World development, including education, health care and global nutrition (United Nations Development Program 2001). They were intended to represent a renewed assault on global inequality to herald the new century. Their erosion will exacerbate global inequality. Climate change is also likely to exacerbate inequality, particularly through its unequal impact upon food production (McMichael & Butler, submitted).

Why is there a denial about limits?

Economists and evolutionary theorists know that resources are always limited. There is almost universal acknowledgement that the population size of non-human species is restricted by a balance between resources, competitors and predators. There is experimental evidence that non-human species behave in ways that suggest a comprehension of limits. For example, experimenters manipulated the nest sizes of a bird called the collared flycatcher.

They concluded that the nesting decisions of these birds were in part determined by knowledge (a form of public information) of their neighbours' reproductive success. Where eggs were plentiful, settlement was increased. But if the hatchlings were scrawny, naïve birds were more likely to go elsewhere. Moreover, evidence suggested that the conditioning of the birds influenced their tolerance of hardship. Birds reared in an artificially impoverished area were more likely to put up with such conditions, whereas birds more familiar with abundance were more likely to reject settlement in over-populated areas (Withgott 2002).

Yet, few mainstream demographers and economists recognise that the concept of human carrying capacity is meaningful (McMichael, Butler *et al.* 2003). This is despite the recognition by economists that scarcity is a fundamental characteristic of society. The extreme form of this disconnection by some academic disciplines with physical reality can be caricatured as the 'cornucopian enchantment' – the idea that in an information-based, post-material economy sufficient goods and services can be created in which the wants of an almost infinitely sized population can be satisfied.

Such an idea is absurd. The global cake is large and has, in the last two centuries, been greatly increased by human ingenuity, technology and cheap energy. Predators and most competitors for humans have been almost eliminated, apart from *other* members of our species. Many people receive only crumbs from this large cake, and most influential economists, trained by and loyal to the dominant power structure, urge that recent policies be intensified.

'Star Trek'-inspired futures of universal abundance are likely to remain elusive, and even though the percentage of malnourished people is reported to have fallen, the absolute number of macro and micro-nutrient deficient people alive today exceeds that of the *entire* global population a century ago (Food and Agriculture Organisation 2002).

Political instability, corruption, poor governance, and terrorism have a complex relationship with scarcity. It is striking how often resource-scarcity and inequality are overlooked as fundamental drivers of human conflict. Instead, it is often argued that sufficient resources and ingenuity exist, regionally and globally, to eliminate conflict and to generate 'wealth for all'. This argument leads to suggestions that the failure to achieve peace is primarily because of the irresponsible behaviour of political leaders.

In reality, limited environmental and human resources are important drivers of conflict and poor governance. This is particularly clear in many parts of Africa and the Middle East. Most proxy conflicts of the Cold War also concerned the control of resources, such as oil in Indonesia and minerals in the Congo. Religious and ethnic conflicts are usually also compounded by differences, real and perceived, in access to resources such as water, land, education and employment.

Limited resources and the still rapidly growing global population are also factors that underlie the global war on terror. More recently, the US and her allies have repeatedly denied that control of Iraq's vast oil resources is even a minor consideration for occupation of that country. This denial is untenable. The alternative explanations for the 2003 invasion of Iraq are insufficient to explain the magnitude of human life and materials that the US and her allies have invested in the war and occupation. Numerous despots, some of them US allies, have both histories and futures of gruesome human rights abuses, far larger in scale than that of Saddam Hussein and Iraq. Weapons of mass destruction are possessed by many nations, but most are members of, or allies with, the coalition of the willing that invaded Iraq. On the other hand, the Western addiction to cheap oil is undeniable, while the increasing instability in Saudi Arabia, fanned by Al Qaeda, provides a powerful motive to reduce Western dependency on Saudi oil.

It is conceivable that the US has a genuine desire to foster democracy within Iraq, because a democratic Iraq may be perceived as enabling greater long-term access by the US to Middle Eastern oil. This could also lead to more independence from Saudi oil. Unfortunately, the allied strategy to foster such a democracy in Iraq over such a short period is likely to fail.

Struggle between and within species – including humans – will always occur. Peace, therefore, will always be fragile. Human ingenuity has dramatically increased global human carrying capacity, by means such as agriculture, refrigerators, transport, trade and credit systems. Cooperation and competition have both been instrumental in this struggle. However, the larger size of cake that humanity can now collectively extract from the global environment has not eliminated disputes over how it should be divided. The human population has also increased, as have human expectations, along with the means of powerful populations to appropriate resources from less powerful groups. And, even if the human population was stable in size, such disputes would continue, though perhaps with less menace than the near future seems to harbour.

The limits to increasing inequality as a defensive measure

Although rarely stated explicitly, the conduct of many comparatively privileged groups, including many states and alliances, reflects a view that the best way to ensure security against a threatening and resentful outside force is to *intensify* inequality. This strategy holds whether between palace and commoner, gated community and slum, or nation and enemy. The strategy is ancient, plausible and often rational, over the short term. It buys time. Yet, at repeated scales and different times, excessive inequality has proven a poor strategy for wealthy individuals and populations.

Attempting to improve security by increasing inequality is limited. Eventually, it will provoke opposition, resentment, and when circumstances allow, hatred, resistance and attack. It also incurs tremendous opportunity and transaction costs, as evidenced in the current war on terror through security, travel restrictions, and anxiety. Eventually it must become rational to *reduce* inequality. Yet this too incurs risks, especially if residual resentment remains high.

Excessive inequality does more than increase the risk of revolution and terrorism. It also erodes sustainability by widening the separation between decision-makers and the adverse effects that their decisions hold for poor and powerless populations. What is important is a greater degree of *global* democracy, so that wealthy populations become more accountable for the adverse effects that their decisions cause. For example, decision-makers would be far more likely to reduce fossil fuel consumption if they knew that they and their descendants had to live on a low-lying Pacific Island, at risk of drowning because of climate change.

Conclusion

The World Scientists' 'Call for Action' signed by over 110 Nobel Prize winners (Union of Concerned Scientists 1997) lamented the 'woefully inadequate' progress made in response to the World Scientist's Warning to Humanity, made in 1992. This earlier warning argued that

> ***… human beings and the natural world are on a collision course … If not checked, many of our current practices put at serious risk the future that we wish for human society … Fundamental changes are urgent if we are to avoid the collision our present course will bring about.***

It also prophesised that '... no nation can escape from conflicts over increasingly scarce resources'.

Einstein warned that the atomic bomb necessitated new ways of thinking if we are to avoid unparalleled catastrophe. The confluence of weapons of mass destruction with the increasing willingness of young people to violently sacrifice themselves in attacks aimed at populations perceived as excessively privileged and indifferent means, at the least, a highly uncertain future for civilisation.

The nuclear brinkmanship of the Cold War is being replaced by a less defined but equally intractable threat. Ahead, climate change, ecosystem degradation, and continuing population growth are likely to exacerbate resource scarcity, thus increasing the potential for violent conflict. Unpalatable as contemplation of this future is, the best hope for civilisation lies in openly confronting these challenges. Facing up to the reality and limits of carrying capacity, and the risk of inequality to *privileged* as well as impoverished populations appears deeply threatening to political and economic elites. The intensity of this denial is intriguing. Although military and political strategists may have cogent reasons for the suppression of discussion of these matters they cannot deny that humanity is now in a precarious state.

Australia is an island, but its sustainability depends upon many international phenomena, both social and environmental. Though contested, the emergence of global terrorism is a plausible response to the perception and reality of excessive global inequality. Terrorism threatens to unravel the sustainability transition by destroying infrastructure, eroding trust and its numerous opportunity costs. The current response to terrorism threatens to generate a 'fortress world', in which bunkers of good governance and prosperity huddle against an increasingly lawless, hostile and barbarous outside.

Globalisation means that it is increasingly meaningful to analyse the world as a linked, interdependent economic, political and strategic unit. This means that, unlike one hundred years ago, poverty in the back blocks of Afghanistan, Nigeria and the Solomon Islands has widespread strategic as well as humanitarian significance. This world, awash with resentment, maldistribution, and weapons of mass destruction requires a clearer discussion of the root causes of conflict to avoid calamity. Military conflict threatens to undermine sustainable development, and is also more likely to occur as a consequence of approaching environmental limits, whether manifest as oil shortages, regional food scarcity because of climate change, or a shortage of well-watered, fertile soil.

Open consideration of carrying capacity and inequality, rather than aggravating instability as many strategists appear to think, may instead be a vital element towards developing a more mature and sustainable civilisation.

References

Bales, K. (1999). *Disposable People: New Slavery in the Global Economy.* Berkeley, Los Angeles, London, University of California Press.

Beder, S. (1998). *Global Spin: The Corporate Assault on Environmentalism.* Melbourne, Scribe Publications.

Boehm, C. (1999). *Hierarchy in the Forest: The Evolution of Egalitarian Behaviour.* Harvard University Press.

Bowles, S. & Gintis, H. (2002). Homo reciprocans. *Nature* **415(6868):** 125–8.

Butler, C.D. (2000). Inequality, global change and the sustainability of civilisation. *Global Change and Human Health* **1(2):** 156–72.

Food and Agriculture Organisation (2002). *The State of Food Insecurity in the World 2002.* Rome, Food and Agriculture Organization of the United Nations.

Friedman, T.L. (2004). The war of ideas. *New York Times.* Parts 1–6, 8–25 January 2004.

Krugman, P. (2002). For richer. *New York Times.* 20 October 2002.

Lovelock, J. (1988). *The Ages of Gaia. A Biography of Our Living Earth.* Oxford, Oxford University Press.

McMichael, A.J. & Butler, C.D. (submitted). Environmental change and food production: Consequences for human nutrition and health. *Asia Pacific Journal of Clinical Nutrition.*

McMichael, A.J., Butler, C.D., *et al.* (2003). New visions for addressing sustainability. *Science* **302**: 1919–20.

Miller, J.H. & Page, S.E. (2004). The standing ovation problem. *Complexity* **9**(5): 8–16.

Nowak, M.A. & Sigmund, K. (2000). Shrewd investments. *Science* **288**: 819–20.

Rogers, P. (2000). *Losing Control : Global Security in the Twenty-first Century.* London, Pluto.

Union of Concerned Scientists. (1997). *World Scientists' Call for Action.* Cambridge, Mass., Union of Concerned Scientists.

United Nations Development Program. (2001). *Human Development Report 2001. Making new technologies work for human development.* New York, Oxford University Press.

Wade, R.H. (2004). Is globalization reducing poverty and inequality? *World Development* **32**(4): 567–89.

Waldrop, M.M. (1992). *Complexity. The Emerging Science at the Edge of Order and Chaos.* London, Viking.

Withgott, J. (2002). Birds spy on neighbors to choose nest sites. *Science* **297**: 1107–8.

Wright, R. (2000). *Non Zero. The Logic of Human Destiny.* New York, Pantheon Books.

4 – The transition to a post-growth society

Clive Hamilton and Richard Denniss

The Australian economy has grown strongly for the last 50 years. What we now need to do is concentrate on the balance between the economy, the well-being of the community and the sustainability of the natural environment. A reduction in the rate of economic growth would not imply that our standard of living would decline. On the contrary, abandonment of economic growth as an end in itself is a prerequisite to a sustainable future Australia. In the words of John Stuart Mill 'a stationary condition of capital and population implies no stationary state of human improvement ... there would be as much scope as ever for all kinds of mental culture and moral and social progress; as much room for improving the Art of living, and much more likelihood of its being improved, when minds ceased to be engrossed by the Art of getting on.'

Economic growth refers to expansion in the volume of goods and services produced by a country. It is typically measured by the annual rate of increase of Gross Domestic Product (GDP), that is, the value of final goods and services produced each year. The economy is said to have grown by 4 per cent in a year when the value of goods and services included in GDP expands by 4 per cent. Economic growth has the potential to solve many of the problems faced by individuals, regions and countries.

The potential for growth to solve problems should not, however, be confused with an inevitability that it will solve problems. Nor can economic growth solve problems that are not economic in nature; indeed, it may exacerbate some social and environmental problems. The housing industry has been growing rapidly in Australia in recent years, but new houses are not being built to house the homeless. The rate of economic growth is also closely related to the rate of growth in greenhouse gas emissions.

The Australian economy grew by more than $20 billion dollars during 2003. That is, at the end of 2003 as a nation we had an additional $20 billion to allocate towards our most pressing problems. On the surface such a fortunate turn of events would appear to provide a good opportunity to address the shortages of resources available for health care, education and environmental protection. Unfortunately, this was not to be.

In Australia, as with most market economies, economic growth is directionless. As more and more resources become available they are not necessarily channelled to where they are needed most. In fact, they are just as likely to be allocated to new 'problems', such as the 'shortage' of mobile phones or the 'shortage' of flat-screen televisions, as they are to old problems, such as the shortage of teachers, nurses or public transport infrastructure. Governments can choose to direct economic growth to where it is most needed or they can choose to distribute it to individuals to fund increased consumption expenditure on whatever consumers prefer.

The costs of pursuing economic growth

While much has been written about the harmful effects of past economic growth on the environment, from a policy point of view it is more important to focus on the potentially harmful impact of the continued pursuit of economic growth.

The important policy question is related to how we can change our economic systems and our lifestyles to improve the balance between the economy and the natural environment. In attempting to answer this question, the policy process is constrained by the presumption that economic growth must always take priority.

Proposals to shift to cleaner forms of energy, reduce industrial pollution or replace increased road construction with greater investment in public transport are usually criticised on the basis of their adverse effects on economic growth. This criticism usually comes from those who benefit from the unquestioned pursuit of more economic growth. Even if it were true that a shift from relying on fossil fuels to a greater reliance on renewable energy would result in reduced economic growth, it is implicitly assumed that any environmental benefits associated with the use of clean fuels would be of less value than the increased consumption expenditure forgone.

It is important to note that declines in the rate of economic growth, such as those typically assumed to accompany a shift to renewable energy or reduced pollution, do not imply that our standard of living would also decline. A slower rate of growth simply means that the rate of increase in the size of the economy would be lower; it does not mean that the absolute level of economic growth would decline. Yet the claim that any particular policy proposal designed to improve the environment, or people's quality of life will reduce economic growth is, in contemporary political debate, a killer blow.

Consider the list of environmentally desirable policies which, after years of discussion, are yet to be implemented for fear that they will 'reduce the rate of growth':

- signing the Kyoto Protocol
- implementing a carbon tax
- ending land clearing
- placing a levy on plastic shopping bags
- making manufacturers responsible for the disposal of their products
- charging agricultural and industrial users for the full cost of the water they use
- ending old-growth logging, and
- abolishing the import duty concessions that ensure that large 4WDs pay a lower rate of tax than small passenger vehicles.

Governments have been aware of the need to implement such straightforward measures for many years. For example, water continues to be used unsustainably in Australian agriculture and industry despite decades of evidence of the need for reform. Yet, despite the widespread support for 'sustainability', in a contest between a better environment or a bigger economy the latter seems to win each time.

While Australia's failure to improve its environment in several crucial areas is lamentable, the pace at which we have gone backwards in many policy areas is even more alarming. In recent years, many policies have been introduced that are unambiguously harmful to the natural environment. In relation to transport, for example, the Federal Government has:

- reduced the price of petrol by abolishing the indexation of fuel excise
- reduced the price of petrol for business users by a further nine cents per litre with the introduction of the GST

- imposed a 10 per cent tax (the GST) on public transport fares, and
- extended the diesel fuel excise rebate.

All of these increase the amount of fossil fuels burnt, with the associated increases in urban air pollution and greenhouse gas emissions. They also shift the relative costs of individual transport modes for consumers. There is nothing 'sustainable' about increasing the financial incentives to shift from reliance on public transport to heavier reliance on private motor vehicles.

In contemporary debate it is rare to find a policy maker who will argue against protecting the natural environment. However, it is also rare to find a senior policy maker who will agree that protecting the environment is more important than maximising the rate of economic growth. Most will agree that both are important, but priorities are revealed in actions not in words. There have been very few prominent policy decisions in Australia that have explicitly sought to protect the environment in the full knowledge that such protection would be likely to reduce economic growth. When environmental protection does cause harm to an industry, such as reducing access by fishers to the Great Barrier Reef, policy makers typically stress the economic benefits that will accrue to other sectors of the economy (for example, tourism).

The pursuit of economic growth stands in the way of solving a wide range of environmental and social problems. Talk of 'win-win' policy outcomes, where both the economy and the environment benefit, typically masks an asymmetry in the policy-making process: policies that are good for the economy but bad for the environment are allowed but policies that are good for the environment but bad for the economy are not. In modern Australia, social and economic problems will remain unsolved unless it is possible to develop policy responses that address those problems without threatening to reduce the rate of economic growth.

Is a post-growth society possible?

> *It is scarcely necessary to remark that a stationary condition of capital and population implies no stationary state of human improvement. There would be as much scope as ever for all kinds of mental culture, and moral and social progress; as much room for improving the Art of living, and much more likelihood of its being improved, when minds ceased to be engrossed by the Art of getting on* (John Stuart Mill, *Principles of Political Economy*, 1865).

The power of language should not be ignored in analysing the desirability of economic growth. For a society to be 'growing' implies that it is vital, alive, expanding and achieving new possibilities. To oppose growth seems destructive, backward, and retrograde.

But economic growth is not the only kind of growth that society can experience. Cultural growth, personal growth, scientific advancement, and even spiritual growth are all attainable in a society where the amount of goods and services being produced per capita is stable. In fact, it is highly likely that these other dimensions of our selves and our society would be far more likely to experience rapid growth if our creative energies and our physical and mental efforts were diverted away from the production and consumption of new consumer 'needs'.

Leaving aside the potential benefits of a post-economic growth society – as, no doubt, supporters of the status quo will do – there are wide-ranging problems and pitfalls said to be inevitable if a society seeks to pursue low, or zero, economic growth. There are three main objections that are typically raised against a post-growth society, none of which remain persuasive when carefully considered.

1 Without growth capitalism will collapse and everything will be thrown into chaos

The most trenchant opponents of capitalism join with its most ardent supporters in this view. The argument from the Left derives from Marx's analysis of the accumulation of capital in which continuing expansion is at the very core of 'the logic of the system'. Neoliberals share this belief in the logic of the system and, like their counterparts on the Left, have difficulty going beyond the mere assertion that capitalism must continue to grow or it will die. This chapter does not attempt a detailed critique of Marx's accumulation thesis, but some observations suggest a response.

First, if capitalism must continue to grow, must it grow at 4 per cent per annum or will 2 per cent be enough? Would 1 per cent do, or something less? At what point would capitalism collapse?

It is common for countries going through the industrialisation process to sustain growth rates of 6 or 8 per cent for two or three decades but then see the rate fall to the 2 or 3 per cent that typifies rich countries. In the 1990s the Japanese economy went through an extended recession in which the average growth rate in the decade to 2002 was around 1 per cent. In a country that had been the envy of the Western world and the model for the Asian tigers, growth effectively stopped for a decade. This has led to much soul-searching and numerous dire predictions of Japan's terminal decline. Yet after some years neither the economy nor the society had imploded. Indeed, prominent Japanese economists and commentators began to argue that a zero-growth economy is to be welcomed.

Some Japanese commentators note that, while official unemployment has grown to 5 per cent, the slump has provided an opportunity to change entrenched work practices that have been literally killing people through overwork and to trigger a cultural renaissance that might rescue Japan from the emptiness of its affluence. Economic stagnation has provided the opportunity for Japanese society to slow down and reconsider the emphasis that the Japanese growth machine has given to expansion at the cost of families, communities and the natural environment.

2 *Unless the growth rate is maximised unemployment will increase inexorably*

There is no doubt that unemployment in our societies causes a great deal of misery, but we must question whether maximising the growth rate is the only or the best way to deal with it. In economics, the informal relationship known as 'Okun's law' suggests that the rate of unemployment will fall by 1 per cent if the real economic growth rate reaches 2.5 per cent above the long run average growth rate. If we embrace slower growth rates, and eventually perhaps a stationary state, is this not a recipe for mass unemployment? The relationship between growth of the economy and growth of employment is more complicated than Okun's law allows. For example, the relationship depends heavily on legal, political and cultural institutions as well as the nature and source of economic growth. Economic growth fuelled by increased consumption spending among high income earners will create significantly fewer jobs than if the same amount of economic growth were to be fuelled by government expenditure on nurses and teachers.

Moreover, since the 1970s chronic unemployment in Europe has often seemed intractable even in the presence of strong growth over many years, a situation sometimes referred to as 'jobless growth'. In other words, the relationship between economic growth and employment growth is highly conditional. Certainly, there is no 'desire' on the part of capitalism to create jobs and resolve unemployment. Job growth is an accidental, rather than a central, element of the modern economic system. Indeed, it is often the case that the financial markets react poorly to a reduction in unemployment, sparking a sell-off that helps dampen the economy and shed jobs.

In the longer term, stabilising population size will reduce the pressure to create the new jobs required each year to absorb new entrants into the labour market. Of course, the principal characteristic of a post-economic growth society is that people no longer feel the compulsion to consume as much and, as a result, most will need to devote less time to paid employment. In other words, while a lower growth rate would reduce the demand for labour, falling interest in consumption would mean less labour supplied to the market. This provides the right environment for policies aimed at redistributing work from those who have too much to those who have too little.

Such policies are already being introduced in some countries. In 1998 the French government legislated for a 35-hour week to apply from the beginning of 2000, with no loss of pay and with penalties for employers who fail to comply. Significant numbers of citizens of rich countries have already decided to accept lower levels of material consumption by downshifting (see Hamilton & Mail 2003; Hamilton 2003; Schor 1998), and many more would like to make such a change if the conditions were more conducive (such as labour market policies that encourage leave purchasing and job sharing). Moreover, the level of employment depends heavily on the choices that are made about the mix of capital, resources and labour used in production processes. In a post-growth society where sustainability is taken seriously, a suite of policies, including ecological tax reform, would encourage substitution of labour for natural resource use and, less so, capital.

3 Without the preoccupation with making money the incentives that motivate humans to achieve will evaporate

There is little evidence to sustain this conclusion, and it is more plausible to believe that humans are predisposed to engage in purposeful activity. After all, one of the most corrosive effects of unemployment is the lethargy it generates in some people, and one of the keys to 'successful ageing' is to remain active. Nor should we fall for the hoary argument that capitalism is the inevitable expression of the selfishness and greed of 'human nature'. Selfishness and greed are socially conditioned (and, as it happens, make their bearers miserable). The apologists of market capitalism, including conventional economists, have simply annexed 'human nature' for their own purposes. But even Adam Smith, the Scottish philosopher and economist to whom the neoliberals improperly trace their belief that the systematic pursuit of self-interest is the best way to advance society's interest, had a more subtle and complete understanding of human motivation.

> ***The disposition to admire, and almost to worship, the rich and the powerful, and to despise, or, at least, to neglect persons of poor and mean condition, though necessary both to establish and to maintain the distinction of ranks and the order of society, is, at the same time, the great and most universal cause of the corruption of our moral sentiments.* (Adam Smith, *Theory of Moral Sentiments*, 1759)**

Already in society today there are large numbers of people who have decided to devote themselves to activities other than the pursuit of higher and higher levels of income and consumption. In fact, 23 per cent of Australians aged 30–59 have voluntarily made a decision such as taking a lower paid job or working part-time which reduced their incomes but improved their quality of life, according to recent surveys (Hamilton & Mail 2003).

With time, a post-economic growth society might witness far-reaching changes in the structure of the market economy. For instance, restrictions on manipulative marketing could see the vast resources of the advertising industry redirected to more worthwhile and creative pursuits than convincing people that two brands of cola are significantly different substances. Many economic activities that are essentially parasitic and make no contribution to social welfare, other than the incomes they provide to the corporations in question, could slowly wither away. Large segments of the financial services industry, especially those involved in speculation and tax avoidance, would no longer be supported. And, as people increasingly rejected the false promise that increased incomes and increased consumption result in increased happiness, industries devoted to producing luxury goods would decline.

Where to from here?

In order to shift away from the pursuit of economic growth as an end in itself, significant changes need to occur at both government and individual levels. If in a well-functioning democracy the majority of the adult population desires such a shift, then significant change should not be too difficult to achieve. A range of actions is, however, required. We focus here on the imminent actions governments could take while recognising that achieving a post-economic growth society would represent a far-reaching social transformation.

1 Increased accountability

At the government level there is a need for much greater openness and transparency in the decision-making process. If the current Howard Government believes that undirected economic growth and increased consumer spending is the pre-eminent national goal, then it should be explicit about this objective. However, if this is not the case, the conditions under which it is willing to pursue a policy not directed at maximising the rate of economic growth should be spelled out. Alternatively, in the unlikely event that the Government were to hold the view that economic growth is not an end in itself, that growth is simply the end result of a range of other policies that are desirable in their own right, then it should refrain from referring to the importance of economic growth when it is arguing for or against a particular change in policy stance.

Similarly, opposition political parties need to spell out clearly where they see economic growth in terms of the hierarchy of priorities. As discussed above, attempts to characterise a political party as seeking to prioritise both economic growth and other objectives should be dismissed out of hand. While some may say that they seek to balance such competing objectives, any party that seeks to maximise the rate of growth cannot simultaneously pursue social or environmental policies that, while increasing well-being, result in a lower rate of economic growth.

2 Review of the policy formulation process

In the last two decades there have been numerous reports and inquiries, by all levels of government, which have considered mechanisms to reduce the impact of economic activity on the environment. As discussed above, however, in many areas little progress has been made while in other policy areas the situation has actually deteriorated.

If significant environmental improvements are to occur, then an inquiry into the political obstacles that stand in the way of environmental reform is required. That is, intellectual and creative energy needs to be diverted, at least temporarily, away from *considering* what needs to be done and towards *developing* a strategy to implement the measures already accepted as necessary.

While it is possible to imagine a world in which good policy ideas are implemented once they have been developed and disseminated, this is not a world that resembles the Australian economy. For example, Australians continue to dispose of over seven billion plastic shopping bags each year, nearly 1000 per household per annum. Despite the evidence of the effectiveness of small levies on shopping bags in reducing demand in countries such as Ireland, no such levy has been imposed in Australia. Both the manufacturers of plastic bags and the Australian retail industry are opposed to the introduction of such a levy, their opposition being based on self-interest rather than the evidence of policy effectiveness. It is therefore naïve to assume that more evidence will lead to a change in the attitudes of those opposing the levy.

The successful implementation of significant environmental policy will, no doubt, impose substantial costs on some industries, companies and even individuals. Those who wish to drive large 4WDs that currently have explicit tax concessions, for example, will need to pay higher import duties, higher registration fees and significantly higher fuel bills if the Australian Government introduces environmentally desirable transport policies. Such a prospect will ensure continued opposition from a small number of consumers and manufacturers. If the desire is to achieve change rather than amass further knowledge, there must be a shift in emphasis from inquiries into the environment to inquiries into implementing environmental policies.

Australia needs to develop a comprehensive framework for the implementation of environmental policies. A good model for such a framework can be found in the Hilmer Report and the subsequent development of National Competition Policy (NCP). Governments initiated a national approach to competition policy reform in October 1992 when they established an Independent Committee of Inquiry into a National Competition Policy for Australia. (The committee's recommendations have become known as the Hilmer Report, named after its Chair, Fred Hilmer (see NCC 2004).) The development and implementation of NCP saw the formation of new enforcement agencies (the Australian Competition and Consumer Commission (ACCC) and the National Competition Council (NCC)) as well as the creation of a system of financial rewards and penalties designed to encourage or coerce participation by state and local governments. National Competition Policy has succeeded, on its own terms, despite the trenchant opposition of large numbers of voters, workers, unions, companies and State Governments. It did so, in part, by identifying the likely winners and losers, harnessing the support of the likely winners and compensating the most politically difficult opponents. Such a pragmatic, political approach to environmental policy is essential if significant changes are to occur.

3 *Individual action*

While there is no doubt that many large-scale changes in policy can only be implemented by governments, there is also no doubt that individuals have an essential role to play. At the very least they must demand more information from their political representatives about their environmental priorities and elect those representatives who will do the most to seek the necessary changes. There are, of course, many more direct actions that individuals can take.

Individual choice plays an essential role in the decision-making process of manufacturers and retailers. The consumer decision to demand free-range eggs, for example, played an important role in changing the behaviour of egg producers. That said, individual action must only be seen as the first step towards industry reform as, in many markets, individuals are not the only, or even the major, purchaser of the final product. Eggs are purchased wholesale by bakers and food manufacturers as well and it is unlikely that they will demand improvements in the treatment of chickens if the result is an increase in their costs. Consumer demand can begin the process of industry reform but it is likely that legislation will be required to complete the process.

In addition to considering social and environmental factors when making purchasing decisions, individuals can also effect significant change by rethinking their lifestyles more generally, particularly by rethinking the significance they attach to the pursuit of high levels of income and consumption expenditure.

In broad terms, national product, as measured by GDP, is equal to the sum of all incomes earned in Australia. If individuals worked fewer hours in order to spend more time with family and friends, or participating in artistic, sporting or community activities instead of simply pursuing higher incomes, then the rate of growth of GDP will fall. As discussed above, nearly a quarter of adult Australians have opted for lower incomes as a way to improve their quality of life in the last 10 years (Hamilton & Mail 2003). If, for example, more Australians chose to forgo higher income by substituting increased leisure time for longer work hours then the environmental benefits would be substantial. Similarly, if people with full-time jobs elect to reduce their hours or weeks of work rather than pursue higher incomes, then there would be increased work opportunities for those who are seeking additional hours of work.

Conclusion

The argument here is that we should go beyond growth and focus our personal attention and public policies on those aspects of life that do, in fact, contribute to our well-being. In a wealthy country such as Australia, the pursuit of high rates of economic growth can be put to one side. Such a change would mean that working life, the natural environment and the public sector would no longer be sacrificed in order to push up the rate of growth. This would almost certainly result in the rate of economic growth, as it is now measured, declining. But it does not mean that the Australian standard of living would diminish. On the contrary, the point of policies designed to improve the natural environment, reduce overwork and increase opportunities for community, sporting and artistic engagement is to improve the standard of living. The key to such a shift, however, is to expose the flawed belief, at both individual and government levels, that higher and higher measures of income and consumption will inevitably result in increased well-being.

The Australian economy has grown strongly for the last 50 years. Real incomes are now three times higher than in 1950. However 47 per cent of people in the highest earning households believe that they cannot afford to buy everything that they really need (Hamilton 2003). The only people who benefit from the perpetuation of the belief that additional appliances provide additional happiness are those who own the companies that manufacture, retail and finance them.

Philosophers have considered the question 'What is the meaning of life?' for thousands of years and few would ever pretend to have discovered the definitive answer. In recent years, however, the answer has become increasingly obvious to many. For individuals the answer is 'a pay rise' and for society the answer is 'more economic growth'. Such pursuits may provide purpose for those who seek them but if the only way to be happy is to have the most, the happiness of a few must come at the expense of the majority.

References

Hamilton, C. (2002). *Overconsumption in Australia: the rise of the middle-class battler.* Discussion Paper no. 49. The Australia Institute, Canberra.

Hamilton, C. (2003). *Overconsumption in Britain: A culture of middle-class complaint?* The Australia Institute, Canberra.

Hamilton, C. & Mail, E. (2003). *Downshifting in Australia: A sea-change in the pursuit of happiness.* Discussion Paper no. 50. The Australia Institute, Canberra.

National Competition Council. (2004). 'Overview of National Competition Policy', <http://www.ncc.gov.au/articleZone.asp?articleZoneID=16#Article-87>

Schor, J. (1998). *The Overspent American.* Harper Perennial, New York.

5 – Land use and ecosystems

John Williams and Denis Saunders

Two centuries of European-style industrial agriculture has resulted in degradation of Australia's shallow soils, with increasing salinity, decreased and polluted river flows, loss of biodiversity and declining trade opportunities. Restoring an ecologically sustainable environment will require a nation-wide revolution in land use, involving a mosaic of commercial enterprises in partnership with natural ecosystem services, including the conservation of Australia's native vegetation and cessation of land clearing. Tackling the causes rather than the effects of land degradation will require an integrated approach involving cooperation between agriculturalists, scientists and governments to achieve sustainable food production for Australia and its trading partners.

Australian agriculture has been very successful for over 200 years, producing substantial wealth to support the nation's economic development. However, we are now producing commodities with ever-declining terms of trade and at significant cost to the environment. We see the results in extensive losses of species, changes in ecosystem processes, and the increasing degradation of our land and water resources.

The search for farming systems and land use patterns that do not harm our environment is urgent. In this chapter we will discuss the challenges we face and pathways towards the solutions.

Australia is a unique, ancient, flat and extremely biodiverse continent with a dry but highly variable climate. Our geological history has created a landscape that has accumulated enormous amounts of salts in the land, lakes, streams and groundwater. Native vegetation has adapted well to these unusual conditions but unfortunately, most of our European-style agriculture, pastures and annual crops, are ill suited to this landscape, although well suited to wet, fertile landscapes. The large-scale clearing and change of hydrological regimes have raised watertables and leached salt in the upper layers of the soil. This has resulted in loss of native species, changes in ecosystem processes and the consequent land and water damage is well documented. Much of the degradation is the consequence of agro-ecosystems that leak carbon, water, nutrients and sediments.

Since about 1985, Australian society, as reflected in the emergence of Landcare and the conservation movement, decided it would no longer tolerate the land management styles that resulted in losses of biodiversity and land and water degradation. We now allocate considerable, but inadequate, resources to fixing the problems. In the past 15 years, Landcare and other institutions have produced significant changes in the attitudes and activities of land managers, industries and the community. Land managers, however, must now shift their focus from treating the symptoms to treating the causes of the degradation.

This will not be easy, because the solutions lie in a nationwide revolution in land use. We need to build ecologically sustainable landscapes consisting of a mosaic of commercial land uses that can capture this leakage of carbon, water, nutrients and sediments and turn these into wealth-creating food and fibre products. To create and shape such a future we need to move from producing the familiar commodities to producing new products for new markets that demand goods produced in environmentally benign ways. We need commercial land uses that do not cause further species losses or impacts on ecosystem processes including damage to land and water.

These new products and land uses will need to be coupled with native ecosystems that provide a suite of ecosystem services (such as the production of clean water, maintenance of fresh air, and conservation of healthy soils) that are valued and paid for by stakeholders and beneficiaries, public and private. This will require innovative and inclusive approaches that permit fair comparison of market and non-market values. Developing the concept of valuing and marketing ecosystem services as part of this process will be increasingly important.

Discovering and building new land use practices that meet these essential criteria require solutions to scientific and technical problems that are many, complex and difficult. For example, the deficiencies in our scientific understanding of the ecology of the rehabilitation process in Australian ecosystems and landscapes are such that we risk creating new problems while attempting to solve current ones.

As well as new scientific knowledge and strong economic drivers, solutions to environmental and natural resource issues require institutional, structural, social and economic changes. Therefore we need rural communities working with scientists, sociologists and economists to build new systems, and close cooperation between the many organisations involved in research to underpin the development of ecologically sustainable agriculture.

To build the picture of what is required, we will first discuss the source of the problem by surveying the extent and effects of natural resource degradation, and then ways of addressing the problem. Strategies include:

- conserving native vegetation, which will mitigate degradation by providing ecosystem goods and services and restoring ecosystem function
- applying landscape design principles to vegetation – both remaining patches of native plants and revegetation – to help restore biodiversity and reduce degradation, and
- radically redesigning land use to enable the sustainable management of rural landscapes and to deliver income to support sustainable communities.

Finally, we review the next steps in finding the scientific solutions, including new roles for people and institutions.

Natural resource degradation: extent and trends

To appreciate the urgency and size of the task ahead, we need to understand the cause and extent of the problems and where we are headed if we continue with current practices.

The main forms of degradation in the Australian landscape are:

- loss and fragmentation of habitat, particularly on the more productive soils
- decline of remnant vegetation, including riparian vegetation
- decline of billions of paddock trees
- loss of native species
- decline in native pastures and environmental value of rangelands

- deteriorating soil quality caused by depletion of nutrients, acidification, and structural and biological decline
- dryland and irrigation salinisation of rivers and land
- water and wind erosion
- changes in riverine processes and loss of essential environmental flows
- movement of nutrients, salts and pollutants to rivers, wetlands and water bodies
- contamination of groundwater with nutrients, salt and pollutants, and
- contamination of land with residues of agricultural chemicals.

The scale and some of the effects of different forms of degradation are summarised in Table 5.1.

The effects of degradation are enduring, not easily reversed, and are becoming increasingly expensive to correct. Such damage has:

- reduced the productive capacity of lands (although in some areas productive capacity has increased)
- adversely affected biological diversity and water quality
- put agricultural trade at risk through contamination and failure to demonstrate production systems that do not damage the environment, and
- threatened human health.

If we fail to build sustainable farming systems in an integrated, inclusive and adaptive way, we will see further losses of biodiversity and changes in ecosystem processes, resulting in more land and water degradation and the cessation of agricultural production in the worst affected areas.

How conserving native vegetation can prevent land and water degradation

Arguably, Australia's most serious environmental problem is the loss of biodiversity – plants, animals, micro-organisms, the genes they contain, the ecosystems of which they form part and the interactions between these life forms and the environment. The most severe losses are in Australia's agricultural zones. In many areas within these zones, less than 10 per cent of the original vegetation remains, with the cleared areas used for agricultural production. This

Table 5.1 Degradation facts and figures

Salinity impacts
Remnant vegetation and wetlands • 2 million ha of remnants and replanted perennial vegetation damaged or at risk from rising watertables and increasing salinity • Remnant native vegetation and fauna threatened • Riparian vegetation, critical to stream bank stability and wetland areas, is damaged and under increasing threat • In WA, 80% of the length of rivers and streams are degraded by salinity, half the water bird species have disappeared from the many wetlands that were once fresh or brackish, 450 plant species are threatened with extinction.
Land • About 2.5 million ha affected; could increase to more than 12 million ha • Estimated capital value of land lost to dryland salinity exceeds $700 million.
Water quality • Increasing salt concentrations in most streams and rivers • Salinity levels in the Murrumbidgee River are increasing at rates of 0.8–15% per annum • Stream salinity in the Murray River exceeds World Health Organization guidelines for about 10% of the year.
Road, bridges and urban infrastructure • At risk or damaged. For example, the National Dryland Salinity Program estimated that high watertables in south-western NSW affect about 34% of state roads and 21% of national highways, amounting to damage costs of about $8 million per year.
Soil acidification • Already occurs on more than 90 million ha; of these, production could be seriously affected on 33 million ha.
Wind erosion • A major problem throughout the margins of the Australian cereal cropping zone and large areas of the southern rangelands • The loss of the most valuable topsoil is largely irreversible. It reduces the nutrient levels of the soil and its ability to retain moisture for plant growth • An estimated 2.6 million ha of cropland in Victoria, (2.5% of the Murray–Darling Basin) is potentially subject to wind erosion.
Water-induced erosion • Estimated to affect 1 million ha of cropland in Victoria, most within the Murray–Darling Basin • Affects 4.8 million ha (about 0.5%) of grazing land, most within the Murray–Darling Basin • 15 million ha of cultivation land in NSW and 16.3 million ha of grazing land (14% and 15% respectively) are subject to water and wind erosion.

Table 5.1 Degradation facts and figures (continued)

Water quality and river health Agricultural activities affect water quality and the environment in the following ways: • extraction of water from rivers and groundwater for irrigated agriculture severely affects ecosystems of rivers, wetlands and estuaries. The drainage water that returns from irrigation could be heavily polluted by high loads of salt and agricultural chemicals • irrigation uses rivers and wetlands as storages and conduits, thus distorting river and wetland flow regimes • run-off from rural land carries sediment, nutrients, organic matter and agricultural chemicals • land clearing and irrigation activities cause rising watertables and salinisation of rivers and wetlands.

extensive loss of native vegetation is now resulting in massive changes to ecosystem processes and considerable loss of native species.

It is a worthwhile proposition that retaining and/or increasing the native vegetation cover in a landscape will help to retain the native biota (flora, fauna and micro-organisms) and control most land and water degradation. To establish the basis for this concept, it is important to define ecosystem goods and services and to understand how they are often lost as degradation occurs.

'Ecosystem goods' are the products we harvest from ecosystems. Pastoralism and forestry depend heavily on ecosystem goods: pastoralism in the rangelands is based on forage produced by native plants; forestry depends exclusively on ecosystem goods such as soil, nutrients and water. Extensive agriculture also depends heavily on them, but mainly on goods such as cereal crops, domestic livestock and fruit that originated from outside Australia. The array of goods is increasing as landowners begin to find value in natural products in addition to the traditional livestock and agricultural crop products.

'Ecosystem services' include activities such as removal of carbon dioxide, production of oxygen, disposal of wastes, pollination, regulation of the hydrological cycle and maintenance of nutrient cycling.

Most of the biodiversity in agricultural landscapes occurs in the soil. Farmers know that maintaining soil fertility is the basis of all agriculture. They also know it is expensive to keep applying fertiliser, and that acidification, salinisation, compaction, loss of structure and erosion are serious problems that greatly reduce profitability. A common reason for these forms of soil degradation is loss of the soil biota and thus the loss of 'free' ecosystem services that biodiversity provides. These 'free' services and functions on which agriculture depends include soil formation, nutrient cycling, maintenance of hydrological cycles, and pollination of crops.

Ecosystem services need to be factored into decisions about land use. For example, a patch of remnant vegetation may provide more than aesthetics, shade and shelter. It is removing carbon dioxide, producing oxygen, using water, and may have some role in controlling watertables, and the movement of wind and water over the surface of the land. At present, we tend to exclude these services in cost accounting for agricultural production.

How ecosystems function

The cycles and processes we observe within natural landscapes result from the functions and interactions of the many elements, both living and non-living. These processes, such as the recycling of carbon, nitrogen and phosphorus, the hydrological cycle and the transfer of energy, as well as the slower processes of weathering and erosion, all rely to varying extents on the functional role of plants, animals and microbes. In stable ecosystems, supply and demand balance these cycles so that few resources are either lost or accumulate.

Individual organisms, from bacteria to large trees, function for their own survival and reproduction and also play a functional role in wider ecosystems. For example, in performing the functions necessary for its own survival, an annual grass contributes to processes such as soil stabilisation, nutrient and water cycling, providing habitat for ground dwelling animals and food for millions of other living things such as decomposing bacteria or herbivorous insects. The more biodiverse an ecosystem, the more complex the interactions of the functional roles, producing an overall functioning ecosystem much greater than the sum of its constituent parts. Similarly, reduction in biodiversity can safely be assumed to impair the overall functioning of a natural system and on a scale much greater than the sum of parts removed.

As biodiversity is lost, ecosystems become less complex. This initiates a cascade of events that result in long-lasting changes. Simplified ecosystems become less resilient. They are less able to absorb environmental shocks and disturbances while continuing to maintain their original levels of function, such as rates of growth, transpiration, fixation and uptake of nitrogen. Reducing biodiversity means that there are fewer components to buffer the blows inflicted by drought, fire, exotic species and climate change.

Deteriorating water quality and dryland salinity are direct consequences of loss in biodiversity. The widespread removal of native vegetation has had a dramatic and immediate impact upon ecosystem function, causing major changes in the hydrological cycle as well as changes in surface flow of wind and water. The complex root systems of grasses, shrubs and trees and their symbiotic relationship with soil fungi provide a vast network for recycling and redirecting water and nutrients. With their removal, there may be less buffering and more extreme run-off events. More water flows across the landscape, moving topsoil around, eroding agricultural land and silting watercourses. These have become severe degrading forces.

Landscape redesign: using vegetation to enhance biodiversity and reduce land and water degradation

The sacrifice of biodiversity in agricultural landscapes has cost so much in terms of land and water degradation that we need to rethink and redesign our use of these landscapes in ways that maintain their integrity but still provide for their profitable, sustainable use. Part of the solution lies in restoring crucial elements of biodiversity to the landscape and optimising the ecosystem services the biodiversity provides.

We noted above that a patch of native vegetation could provide many ecosystem services as well as aesthetics, shade and shelter. Both the size of a patch isolated from other patches by large areas of agricultural land, and the arrangement of patches on the landscape, have a profound influence on which species survive and consequently, which ecosystem services are maintained.

The key elements of landscape design for protecting terrestrial biodiversity and ecosystem functions are size, shape, separation/connectivity, species composition, and position.

Design elements are most important for landscapes that have been cleared to less than 70 per cent of native vegetation cover. In areas with less than 30 per cent cover, revegetation must be designed if the return-for-efforts are to be maximised. Where there is between 30 and 70 per cent cover, design can help to avoid thresholds of change that cause both rapid loss of species and change in ecosystem functions.

More research is needed into the effectiveness of conserving patches of habitat as a way of allowing species to persist in highly cleared areas. A landscape strewn with small patches of native vegetation could preserve elements of biodiversity, but the cost of maintaining these patches could be much higher than for larger patches. Furthermore, organisms that can only survive in the habitats provided in large patches would become locally extinct. If we are to maintain species and minimise maintenance costs, it would be expedient to design a landscape with an optimum number of small, medium and large patches.

In landscapes where clearing has reduced native vegetation cover to less than 30 per cent, the emphasis is on revegetation. For revegetation to be efficient and effective, we need to refine our understanding of the different landscape design elements. At present, revegetation involves planting or direct seeding, which is both labour intensive and expensive.

Much research is needed in areas such as:

- where in the landscape should vegetation be placed?
- how should the condition of current vegetation be improved?
- what sort of species and structural complexity is necessary?
- how can regeneration of remnant vegetation be used to enhance revegetation?

These important questions must be resolved if we are to address environmental problems in the landscape.

Landscape design for hydrology and ecosystem function

Revegetation programs have multiple objectives and must therefore be designed for restoring ecosystem function, including hydrology, maintenance of habitat and movement of biota.

Clearing or intensification of land use usually occurs in places with the most productive soils. Native species are lost as habitat is lost or degraded. The outcome for fauna and flora will depend on which habitats are cleared and how different species use the landscape. The effect on species of lost connections between patches on native vegetation will depend on their life history, characteristics and mobility.

Management options for controlling salt delivery to land and water include reversing its delivery by reintroducing perennial deep-rooted trees and shrubs. This will greatly reduce the amounts of water moving beyond the root zone and return recharge rates to levels similar to those under native vegetation before it was cleared to make way for annual crops and pastures. Successfully managing the cause of salinity depends on reducing groundwater recharge.

Priority will be given to replanting deep-rooted perennial plants to reduce the amount of water leaking to watertables and the salinisation of rivers, wetlands and valley bottoms. To achieve the best results for rural communities and the environment, we need to integrate planning to enhance biodiversity and to control recharge.

A sustainable land use might consist of:

- 30 per cent of the area permanently covered in native vegetation, including trees, shrubs and grasses
- 20 per cent covered in deep-rooted trees, shrubs and grasses, planted primarily to control recharge and produce income from grazing and farm forestry
- 30 per cent used intensively for annual crops, and
- 20 per cent used less intensively for mixed grazing and cropping.

The correct balance between different types of land use will vary for different catchments.

The amount of remaining native vegetation determines the extent to which biodiversity and ecosystem functions are intact. A number of studies have analysed the changes in the rate of loss of connections between patches, the rate of loss of species, and lag times associated with random and non-random clearing, and identified thresholds of connectivity.

Landscape diagnosis, targets, and management priorities all depend on the landscape context. The scale of revegetation and its strategic location also need to match the particular groundwater system that is controlling the expression of the salinisation process. We are starting to understand how groundwater flow systems can be applied to catchments and used to inform:

- the extent to which we need to change land use and reduce recharge in order to halt, and maybe reverse, the spread of salinity, and
- the lag times between initiating actions to reduce recharge or intercept saline groundwater and any evidence of responses in groundwater levels or salt delivery.

Both the extent of the changes we must make to land use and the time lag before the changes affect salt discharge to land and water resources are far greater than is widely recognised. It is estimated that after measures to reduce recharge have been implemented, there will be no evidence of groundwater and therefore salt responses at the discharge end of the system for 10 years in local systems and for 50–200 years in large regional groundwater systems.

Landscape redesign – possibilities for farming without harming

If we seek sustainable management of rural landscapes, we need to develop and deploy a suite of novel land uses that are matched to the diverse climate, soils, hydrological conditions and, to some extent, the native biota of the Australian continent. These land uses, in combination, need to deliver water and nutrient leakage rates past the root zone or across the land surface that approach those under natural vegetation. This will require radical change to land use incorporating:

- development of commercially driven tree production systems and/or novel tree species for large areas of the current crop and pasture zones. These would include trees to produce fruits, nuts, oils, pharmaceuticals, bush foods and forestry products such as specialty timbers, charcoal, and biomass energy
- new farming systems made up of novel mixes of all the best current annual and perennial plants, the best agronomy, companion plantings, rotations and combinations
- new forms of cereals, pulses, oilseeds and forages selected or bred for characteristics that substantially reduce deep drainage and nitrogen leakage, and
- new land assessment tools that
  - best locate trees, other perennial plants, high-value annuals, and native vegetation to meet water quantity and quality targets, and biodiversity goals, and
  - facilitate identification and re-assignment of land so that on some parts of the landscape, productivity is greatly enhanced (double yield) and other parts are removed from production to provide a range of ecosystem services and protect the native biota.

To realise this vision, we will need to pioneer the development of new landscapes. These will comprise mosaics of tree crops driven by large-scale industrial markets such as biomass fuels, high-value annual crops, mixed perennial–annual cropping systems, and areas devoted to maintaining those elements of the Australian biota dependent on native vegetation. Devising the optimal placement of these land uses for salinity control, productivity

Table 5.2 New commodities and markets for ecosystem services

Commodity	Share of business (%)	Client
Wheat	40	World market
Wool	15	World market
Timber	10	World and local markets for wide range of products from pulpwood to specialty timber; biomass energy for local energy suppliers to national electricity grid or local liquid fuel distribution.
Carbon credits	7.5	Industrial consumers of fossil fuels, e.g. steel mills
Salinity credits	7.5	Public and private funds/enterprises and individuals driving cost sharing for catchment management
Water supply management	15	Water supply companies
Biodiversity credits	5	Public and private trusts investing in natural heritage

and maintenance of native biota will require a robust understanding of landscape processes and functions, good maps of landscape properties, particularly salt storage and groundwater flow, and an understanding of the distribution and abundance of flora and fauna.

Such a mosaic of ecologically sustainable, commercial land uses could be combined with land uses that provide ecosystem services that are valued and paid for by stakeholders and beneficiaries. The consequence would be that rural enterprises might derive their income from sources other than traditional food and fibre production. For example, they might provide services paid for by either private or public stakeholders and beneficiaries, or in some innovative mix. A possible set of diverse sources of income is set out in Table 5.2.

However, while visions for sustainable landscapes are emerging, many of the components described above do not yet exist. A substantial new research and development effort is needed that tackles the redesign of farming systems and their integration into the landscape as a whole. This needs to combine biophysical and economic studies that deliver:

- novel designs well matched to soil, climate and catchment circumstances, including biodiversity

- on-farm measurement and improved land assessment techniques
- modern genetic improvement techniques, and
- a participatory process that engages all land managers.

Farm forestry, new agricultural production systems and restoration of native vegetation

We have discussed the opportunities provided by farm forestry, new agricultural production systems, and restoring native vegetation to present opportunities to restructure the landscape with vegetation that has a water-use pattern similar to that of the original native vegetation, and the potential to substantially ameliorate existing and impending problems. The possibility of implementing this type of solution is increasing. The expansion of forestry on cleared agricultural land is becoming more attractive in higher rainfall zones. Commercial prospects for traditional grazing are poor, while market prospects for expanding plantation forestry appear to be improving. In addition, there is increasing interest in using the ability of trees to sequester carbon as a means of meeting greenhouse commitments. Farm forestry and agro-forestry for the mid to lower rainfall zones appear to offer attractive options, although a great deal more work in building these new industries is needed.

The use of native flora and fauna may form an increasing part of rural production. Bush foods, native wildflowers, and essential and other oils for pharmaceutical or industrial chemicals are receiving increasing attention. Such diversification in farming enterprises will increase the planting of native vegetation and expand production on those parts of the landscape suited to high-value crops and pastures.

The Redesigning Agriculture for Australian Landscapes Research and Development Program, a joint initiative of Land and Water Australia and CSIRO, is researching how agricultural systems can be redesigned to address a range of sustainability issues. This approach has potential to be applied through:

- selection and plant breeding – for commercial crops, pastures and native plants to manipulate phenology, canopy development, rooting function, distribution and temperature response, and
- rotating, and mixing in space and time, innovative configurations of plants involving crops, pastures, forest and horticultural trees, native plants, bush foods, etc. in alleys, blocks, windbreaks and clusters, over rotations of months or years.

We need more information about the water-using capacity of various types of vegetation and on experimentation with new farming systems that are adapted to the temporal and spatial variability of the Australian climate.

Crop and pasture species have not been designed with the control of natural resource degradation in mind. Most plant breeding programs generally focus on grain yield and quality, pest control and other limitations. Few, if any, have focused on the role of crop and pasture species in controlling deep drainage and nitrogen leakage. Studies have highlighted that the breeding, selection and bioengineering of crops and pastures have considerable potential to help ameliorate dryland salinity and acidification by designing crops and pastures to minimise the leakage of water and nutrients past the root zone.

Developing new farming operations that do not harm the natural resources and environment, while generating enterprise incomes that can support sustainable communities, must be an urgent goal for Australia.

Next steps for science, people and institutions

To understand how Australia's rural industries might move towards ecological sustainability, we first need to identify the scientific and technological issues that must be solved. We have flagged some in this chapter.

The very leaky nature of Australian agro-ecosystems lies at the root of nearly all land and water degradation concerns. To match farming and land use patterns to landscape and ecosystem function, we desperately need new biophysical solutions that can plug leaky systems and capture the water and nutrient for productive purposes. The irony of Australian agriculture is that, while the shortage of both water and nutrients greatly restricts yield, the fundamental cause of both salinity and acidification is the loss of valuable water and nutrient beneath the crop or pasture. A key strategy for science and technology, therefore, is to build productive agro-ecosystems that leak much less water, nutrient and carbon to the landscape in which they are located.

This requires that the soil/plant/animal agro-ecosystems be studied in an integrated way and as part of the larger-scale ecological and hydrological processes that operate over the landscape. Solutions must incorporate these functions at a range of scales, including paddock, hill slope, catchment, whole landscapes and regional basins. The landscape designs will need to integrate sustainable production and maintenance of biodiversity for the catchments and regions. Any revegetation program must have multiple objectives and therefore be designed for restoring ecosystem function: hydrology, nutrient cycling, maintenance of habitat, and movement of biota. These will need to be configured with the aid of emerging knowledge of:

- salt storage, groundwater and surface water flows, river form and function
- biodiversity, including ecosystem function
- biogeochemical process and water quality, and
- carbon sequestration.

There are serious deficiencies in our scientific understanding of the ecology of rehabilitation processes in Australian ecosystems and landscapes. We do not know how to reconstruct them. The first step in our search for an ecologically sustainable agriculture requires that we address agricultural production as an agro-ecosystem that is part of larger-scale ecosystems and landscape processes. Knowledge of how best to revegetate land and implement land uses that are ecologically sustainable and can support viable rural communities is critical to any regional development plan. At present, we risk creating new problems while attempting to solve current ones.

In addition to new scientific knowledge and strong economic drivers, solutions to environmental and natural resource issues require institutional, structural and social change. People from all sectors of the community must be involved with scientists from the earliest stages of a program and through each of its phases: planning, research implementation, monitoring and evaluation. Scientific and technological innovation both on farms and in laboratories will play a fundamental and increasing role in the development of sustainable farming. Such innovation requires a paradigm shift by research institutions, rural communities, funding agencies and governments.

Thus the development of farming systems that do not harm the environment involves:

- rationalisation of resources
- a refocus on farming system research within an ecological framework
- adoption of participatory methods of on-farm research, and
- cooperation between universities, CSIRO and state agencies in research and development that underpins the development of ecologically sustainable agriculture.

A significant feature of the future will encompass rural communities working with biophysical scientists, conservation biologists, sociologists and economists to build new systems. The innovative use of on-farm measurement, coupled with simulation models to design and examine alternative operations in terms of both production and impact on the natural resource, will be an increasingly important tool of discovery.

Few commercial farming and other land use systems can control the causes of land degradation while generating farm incomes that can sustain rural communities. We need to rethink and redesign our use of agricultural landscapes in ways that maintain their biotic integrity but still provide for their profitable, sustainable use.

The challenge is to come up with biodiversity-based solutions for reconstructing the Australian landscape and ensuring a sustainable future for rural and urban communities alike.

Acknowledgments

This chapter draws heavily from the work of Matt Colloff and the authors in the development of an unpublished paper prepared for the workshop, 'Murray–Darling Basin 2051: Setting the vision for long-term biodiversity objectives for the Murray–Darling Basin' held in Canberra on 25 and 26 October 2001. The comments, discussions and suggestions of many of our colleagues in CSIRO, particularly Tom Hatton, Glen Walker, Richard Stirzaker, Brian Keating and Ted Lefroy which have been used extensively in this work are gratefully acknowledged, as is the editorial assistance of Hester Gascoigne of Hester Gascoigne and Associates Pty Ltd. The National Land and Water Resources Audit and particularly Adrian Webb and Warwick McDonald provided access to extensive sets of reports and papers including *Australian Dryland Salinity Assessment 2000.* Any errors of analysis and interpretation and text are, however, the responsibility of the authors.

Further reading

Agriculture for the Australian Environment. (2003). In *Proceedings of the 2002 Australian Academy of Science Fenner Conference on the Environment.* (Eds B.P. Wilson & A. Curtis.) Charles Sturt University, Albury, NSW.

ANZECC Taskforce. (2001). Implications of Salinity for Biodiversity Conservation and Management. ANZECC (Australian and New Zealand Environment and Conservation Council), Canberra.

CSIRO Australia. (1996). Bush foods: growing food from the bush. *Rural Research* **172:** 9–23.

Daily, G. (1997). *Nature's Services: Societal Dependence on Natural Ecosystems.* Island Press, New York.

Daily, G.C. & Walker, B.H. (2000). Seeking the great transition. *Nature* **403**: 243–5.

Dunin, F.X., Williams, J., Verburg, K. & Keating, B. (1999). Can agricultural management emulate natural ecosystems in recharge control in south eastern Australia? *Agroforestry Systems* **45**: 343–64.

Hobbs, R.J. & Saunders, D.A. (eds). (1993). *Reintegrating Fragmented Landscapes: Towards Sustainable Production and Nature Conservation.* Springer-Verlag, New York.

James, C.D. & Saunders, D.A. (2001). *A Framework for Terrestrial Biodiversity Targets in the Murray-Darling Basin.* CSIRO Sustainable Ecosystems and Murray–Darling Basin Commission, Canberra.

Lambeck, R.J. (1999). *Landscape Planning for Biodiversity Conservation in Agricultural Regions – a case study from the wheatbelt of Western Australia.* Biodiversity Technical Paper No. 2. Environment Australia, Commonwealth of Australia, Canberra.

Lefroy, T., Hobbs, R. & Hatton, T. (2000). Effects of Changing Vegetation on Hydrology and Biodiversity. In 'Visions of Future Landscapes'. (Ed. A. Hamblin.) *Proceedings of the Australian Academy of Science 1999 Fenner Conference on the Environment,* 2–5 May, 1999, Canberra, Commonwealth of Australia. Canberra.

LWRRDC/CSIRO. (2000). *Scoping study – opportunities to breed/select/bioengineer species to control deep drainage and nitrogen leakage.* Occasional Paper No 09/00. Land and Water Resources Research and Development Corporation and CSIRO, Canberra.

National Land and Water Resources Audit (NLWRA). (2001). *Australian Dryland Salinity Assessment 2000: extent, impacts, processes, monitoring and management options.* Commonwealth of Australia, Canberra.

Price, P. & Williams, J. (2002). Redesigning agriculture for Australian landscapes R&D program. Review of Phase 1, January 2002. Land and Water Australia, Canberra.

Saunders, D.A. (1996). Does our lack of vision threaten the viability of the reconstruction of disturbed ecosystems? *Pacific Conservation Biology* **2:** 321–6.

Saunders, D.A., Hobbs, R.J. & Ehrlich, P.R. (eds). (1993). *Nature Conservation 3: Reconstruction of fragmented ecosystems, global and regional perspectives.* Surrey Beatty & Sons, Chipping Norton.

State of the Environment Advisory Coucil. (1996). *Australia: State of the Environment 1996.* Commonwealth of Australia and CSIRO Publishing, Collingwood.

Stirzaker, R., Lefroy, T., Keating, B. & Williams, J. (2000). *A revolution in land use: emerging land use systems for managing dryland salinity.* CSIRO Land and Water, Canberra.

Walker, G., Gilfedder, M. & Williams, J. (1999). *Effectiveness of current farming systems in the control of dryland salinity.* CSIRO Land and Water, Canberra.

Williams, J. (1991). Search for sustainability: agriculture and its place in the natural ecosystem. *Agricultural Science* **4:** 32–9.

Williams, J. (1995). Farming without harming: How Australia made rural industries sustainable. In *Challenge to Change: Australia in 2020.* (Eds R. Eckersley & K. Jeans.) pp. 223–240. CSIRO, East Melbourne.

Williams, J. (1999). Biophysical aspects of natural resource management. In *Commodity Markets and Resource Management, Proceedings National Agricultural and Resources Outlook Conference, Canberra, 17–18 March 1999*. Vol. 1, pp. 113–23.

Williams, J. & Hook, R.A. (1992). Search for ecological sustainability in Australian agriculture. In: *Proceedings International Conference on Sustainable Land Management, Hawke's Bay, New Zealand, 18–23 November 1991*. (Ed. P Henriques.) pp. 434–48. International Pacific College, New Zealand.

Williams, J., Walker, G.R. & Hatton, T.J. (2002). Dryland salinisation: A challenge for land and water management in the Australian landscape. In *Agriculture, Hydrology and Water Quality*. (Eds P.E. Haygarth & F.E. Jarvis.) pp. 457–475. CAB International, London.

6 – Water: the key to sustainability in a dry land

Peter Cullen

Australia is the most arid inhabited continent, with a low and variable rainfall. The Murray–Darling river system is our most important water resource, comprising 14 per cent of the Australian land mass. Over the past 80 years, extensive use of this water resource, mainly through irrigation and the building of levees, has produced adverse environmental consequences such as reduced river flows, water pollution, downstream salinity and loss of biodiversity. The National Water Initiative proposed by the Council of Australian Governments in 2003 made various recommendations to address these problems. They include improved security of water access, ensuring ecosystem health, an efficient market so water goes to the best use, and conservation and recycling of water in cities. A sustainable water future will entail extensive collaboration between governments and stakeholders to ensure that the true costs of water use are borne equitably and accountably in both rural and urban areas.

Water is the key to living and to economic development in Australia. It is a critical driver for the variety of ecosystems that provide a global 'hotspot' for biodiversity. Water is a State responsibility in Australia, although the Federal Government has some interests and has been a substantial source of funding for water resource development. Australian State and Federal Governments have recently committed to a $1.4 billion joint program (National Action Plan for Salinity and Water Quality) to address issues of salinity and water quality, as a response to the degradation and the risks of salinisation to land and water resources.

Most States have now moved beyond the sort of simplistic water planning that focused on 'yield' in terms of extracting water for irrigation. It is now largely understood that rivers are ecological systems that need to have some water left in them for ecological purposes. This understanding has come about only after the loss of native fish and the occurrence of extensive blue-green algal blooms that have reduced the amenity and utility of many waterways.

We are currently going through a turbulent and exciting period of water reform. The recent decision by The Council of Australia Governments (COAG) to create a national water market opens the way for important new policy opportunities and possible rescue of some of our deteriorating waterways.

The Murray–Darling Basin comprises some 14 per cent of the Australian landmass, about 1 million square kilometres, and yet has only 6 per cent of Australia's run-off. It is highly developed for agriculture, supporting about 42 per cent of all Australian farms. The history of water development in the Murray–Darling Basin is a history of articulate interest groups seeking to have the waters used for their particular advantage. There has always been a tension between the upstream States and the downstream State who had differing views of the public good. These tensions continue, although the issues now relate to over-extraction of water and salinisation of land and water.

Over some 80 years there has been extensive development of water resources of the Murray–Darling, and a highly productive irrigation industry has been developed. Unfortunately, much of this development cannot pass economic nor environmental scrutiny, and there has been considerable degradation.

We have now learned that many of our water management strategies have caused great damage to the ecosystem of the Murray–Darling. By trapping winter stream flow we have starved the river of flow in winter and turned it into a bank-full irrigation channel in summer, totally reversing the natural flow pattern. We have extracted excessive amounts of water and built weirs to help distribute the water. These also turned out to provide ideal habitat for both blue-green algae and for carp. By our water extraction and our building of levees to control floods we have isolated the river from its floodplain. This has had massive impacts on populations of native fish and birds as well as on the River Red Gum forests of the Basin.

Australia is an old and highly weathered landscape, and much of it is very flat. Soluble salts originate from the weathering of rock minerals and the deposition of oceanic salt through rain or wind, and are redistributed in the soil profile and landscape with movements of water. Salinity problems are most common in semi-arid areas where there is sufficient rainfall to wash

Table 6.1 Projections of average salinity levels in selected rivers of the Murray-Darling Basin, in electrical conductivity (EC) units.*

River valley	1998	2020	2050
Murray River at Morgan	570	670	790
Murray River at Renmark	400	480	550
Murray River at Swan Hill	270	270	310
Murrumbidgee	250	320	350
Avoca	970	980	1480
Loddon	870	880	900
Goulburn-Broken	130	180	260
Lachlan	530	780	1150
Macquarie	620	1290	1730
Namoi	680	1050	1280
Condamine-Balonne	210	1040	1040

*Electrical conductivity of water is widely used to measure salinity. The World Health Organization suggests a level of 800 EC units as the upper limit for human drinking water, and 1500 EC units as the threshold for irrigation and environmental purposes.

Source: *The Salinity Audit of the Murray–Darling Basin* (1999).

the salt down a metre or so, but not enough to wash it out of the profile. The salt can then be brought back to the root zone by rising groundwater.

In irrigation areas where excess water has been applied, the groundwater has risen causing waterlogging, commonly bringing salt to the surface. Engineering approaches to this problem have concentrated on providing drainage to remove the symptoms of the problem rather than addressing the causes (an example is the large evaporating basins that have been created from former wetlands).

Native Australian vegetation typically has deep root systems that can extract water from deep in the soil profile. This is an adaptation to the drought periods common in Australia. When deep-rooted vegetation is replaced by shallow-rooted pastures or crops that do not use the rainfall that falls, the watertable often rises, mobilising salt if it is in the profile. It is a paradox that, in a dry country like Australia, a common form of land degradation is due to excess water causing waterlogging and salinisation.

The Salinity Audit of the Murray–Darling Basin (1999) showed high salinity levels in many of the tributaries of the Basin. The levels of salt in the Border

Rivers, Bogan, Condamine-Balonne, Macquarie, Namoi and Warrego Rivers are expected to exceed World Health Organization drinking levels within 20 years, and the Lachlan and Castlereagh rivers within 50 years (see Table 6.1). The Avoca and Loddon Rivers in Victoria already exceed this level.

Irrigation in Australia

Irrigation has long been seen as the key to unlocking the wealth of the Australian landscape, and myths about making the desert bloom abound. In the 20th century there was a view that any water flowing to the sea was wasted, and all we had to do was build massive engineering works to harness the water. These 20th century myths persist amongst some older and influential Australians.

During the current drought, talkback radio hosts in Sydney started calling for massive engineering works to droughtproof our country. This led to the formation of the Wentworth Group of Concerned Scientists who answered these calls with the response that we would do better if we learned to live *with* our country and stop our futile attempts at taming it.

Irrigation in Australia uses around 75 per cent of all the water harvested and was responsible for agricultural production worth $9618 million in 2000–01. In fact, half of the profit from all Australian agriculture comes from the 0.5 per cent of our land that we irrigate. There has, however, been a dramatic increase in irrigation, and water use has increased by 59 per cent between 1983–1996. The competition for water is becoming more intense, just as the community is starting to appreciate that we have degraded the health of our rivers by over-extraction. However, much of this water is used on crops that produce relatively low gross returns (see Table 6.2), and so it may be possible for water to be better used.

The challenge of assessing river health

Our understanding of how to assess river health has developed rapidly over the last decade. Prior to that, most assessments were based on water quality measurements of particular substances in the water. The problem was that these substances varied over often quite short time intervals, and integrating the overall effect was difficult, especially when only sporadic measurements were taken.

In the 1990s it was realised that the biology of the river system gave a far better measure of river health than water quality measurements, since the

Table 6.2 Water use efficiency in Australian irrigated agriculture

Industry	$/ML return
Pasture, grains*	270
Dairy*	529
Rice	179
Cotton	420
Sugar	217
Fruit	1590
Grapes	1859
Vegetables	1817

* These two categories use 50% of irrigation water.

Source: Derived from data on water usage and economic returns included in *Water Account Australia* (2004).

biota observed were a function of the chemical and physical environment experienced by the organism over some particular period of time.

The National Land and Water Audit (2002) reported a biotic index for many Australian rivers based on the observed communities of aquatic invertebrates. These give a consistent and useful measure of the health of the river in comparison to an unimpacted reference site.

Norris (2001) reported a snapshot of river health in the Murray–Darling, based upon aquatic invertebrate populations that showed widespread impacts on the river systems. Other biological measures of the degraded condition of the Murray River include:

- loss of red gums on floodplain below Euston
- listing of Murray Cod as threatened
- decline in native fish and explosion of populations of carp
- cold water pollution extending up to 300 km downstream of major storages
- almost permanent algal blooms in the Lower Murray
- closing of Murray mouth and Coorong.

Despite this growing body of evidence, there are still various interest groups in denial about the health of the Murray–Darling river system. Some upstream irrigators assert the river is in good condition, and there is no need for them to change their irrigation practices. There are good local catches of

Murray Cod, but the populations have dwindled and now appear to be isolated communities. There are no reputable ecologists saying the river is in good health, and those that have years of experience on the river are greatly concerned about its health. Even returning 1500 GL to the river gives only a moderate chance of giving us a healthy river, but unless we do this, we can expect it to deteriorate further.

The water reform agenda of 1994

The reforms started in 1994 when the then Prime Minister and State Premiers agreed, under the auspices of the Council of Australian Governments (COAG), to a series of major reforms that were designed to develop a more competitive water industry and, through annual payments from the National Competition Council, provide significant financial rewards for States that achieved various changes.

The 1994 COAG water reforms encompassed natural resource management, pricing, rigorous approaches to future investment, water trading, allocations of water to the environment, institutional reform, environmental improvements and improved public consultation. Some of these reforms have been implemented but, in general, achievements have so far been modest (see Table 6.3). And while some of the *economic* benefits of these reforms have been achieved, the *environmental* benefits are so far much less obvious.

COAG emphasised the importance of addressing both efficiency and sustainability in water services and water resource management. The reforms cover many aspects of the water industry, including institutional arrangements for regulation, management and service provision, water allocations and entitlements, water pricing, environmental protection and community input into decision-making. The main principles in the reforms as they affect environmental outcomes include:

- pricing based on the principles of consumption-based pricing, full-cost recovery and transparency or removal of cross-subsidies
- future investment in new schemes, or extensions to existing schemes, to be undertaken only after appraisal indicates it is economically viable and ecologically sustainable
- comprehensive systems of water allocations or entitlements, backed by separation of water property rights from land title and clear

Table 6.3 Achievements of the COAG water reforms

Reform element	Outcome
Pricing	Full cost charging for water and wastewater services is now well advanced in the urban sector. Reform in the rural sector has been limited, with few jurisdictions charging the full cost of supply and no consideration for externalities imposed upon others.
Investment in water resources viable & sustainable	There has been little investment in new dams, partly because of previous over-investment but partly due to this requirement for investment to be viable and sustainable. Private capital has not been forthcoming for new dam developments for rural water and these appear unlikely without public subsidy.
Comprehensive systems of water entitlements	Progress in clarifying entitlements to water access has been slow in some jurisdictions due to concerns of States that they may be liable for compensation if perceived rights are removed. Victoria has given irrigators permanent rights under its Bulk Water Entitlement provision, other jurisdictions are allowing for periodic review of access rights.
Water allocations for environment	Progress has been limited. Some allocations have been made for major wetlands. Many rivers are over-allocated and progress in recovering water from irrigation for the environment has been slow. Weak planning frameworks and insufficient knowledge have hampered progress, as has the lack of political will.
Trading	Failures to clarify access entitlements have limited trading, and restrictions to the market have slowed the movement of water from low economic to high economic uses. Trading has led to the activation of unused licenses and increased water extraction, and made costs of recovering water for the environment much higher.
Integrated catchment management	Good progress has been made in developing regional catchment organisations in Victoria and SA, less so elsewhere.
Consultation	Consultative processes of various efficacy exist in all jurisdictions, however, it is often thought that irrigators have a greater involvement than conservation or Indigenous interests.
Research	Although some research has been funded, there is no adequate funding mechanism or organisational structure. Needs a levy on water users to provide funds and requires more effective mechanisms to deliver research findings to those who need them.

specification of entitlements in terms of ownership, volume, reliability, transferability and, where appropriate, quality

- formal determination of water allocations or entitlements, including allocations for the environment as a legitimate user of water.

Environmental requirements should be determined on the best scientific information available and have regard to inter-temporal and inter-spatial water needs required to maintain the health and viability of river systems and groundwater basins.

- trading, including cross-border sales, of water allocations or entitlements, within the social, physical and ecological constraints of catchments
- administration and decision making to provide an integrated catchment management approach to water resource management and establishing arrangements to consult with the representatives of local government and the wider community in individual catchments
- consultation where change and/or new initiatives are contemplated involving water resources
- appropriate water-related research necessary to progress implementation of the framework, including consistent methodologies for determining environmental flow requirements.

While this was an exciting agenda for change that was widely regarded as necessary, even if painful, all States have undergone extensive change as a result, but in reality some of the important outcomes have not been achieved.

The COAG National Water Initiative of 2003

In August 2003, the COAG again considered further water reforms, and committed to a new National Initiative (www.pmc.gov.au/docs/national_water_initiative) with the following goals:

- improve security of water access entitlement by clear assignment of risk and returning over-allocated systems to sustainable allocations
- ensure ecosystem health through protecting ecosystem assets
- ensure water goes to best use through an efficient market
- encourage conservation and recycling in cities.

The outcomes sought from this further initiative was to deliver best practice water pricing, effective management of environmental water, improved monitoring and information including water accounts and urban water reuse and recycling, more efficient technologies and review of pricing.

This was more than just an important and overdue recommitment to the 1994 water reforms, but provided $500 million to recover water for the environment and provided a clear framework to proceed. There is, however, much yet to be achieved. Whether these important outcomes will be achieved or the funds frittered away in endless wrangling, disagreement and the building of further infrastructure desired by the engineers is yet to be seen. The Murray–Darling Basin Ministerial Council has now taken its first step in restoring the health of the Murray River by committing to 500 GL over a five-year period. The amount of water to be recovered for the river comes to about 3.5 per cent of the total water extracted from the river.

Emerging principles to guide water reform

The Wentworth Group released its Blueprint for a National Water Plan in mid-2003, and established some broad principles on which to go forward:

- all Australians have a right to a supply of safe water for domestic use
- all have a responsibility to use water efficiently
- environmental health provides the essential foundation for all other uses. Without healthy rivers other uses by society are not possible
- those who use water to create wealth need security of water supply and must take responsibility for sustainable outcomes
- Australians must become water literate and understand how their individual actions affect our water resources and the resources of those living 'downstream'.

The challenge of how to recover water for the environment in an equitable and efficient way has delayed the returning of water to the environment. The Government started a consultative process on the 'Living Murray program' (see www.thelivingmurray.mdbc.gov.au), but gave no indication as to how it expected to recover the water. It was hardly surprising that this uncertainty caused great concern in rural Australia, and much effort went into challenging the science rather than addressing the problem.

There are four broad strategies for recovering water for the environment:

- take it, by unilaterally reducing allocations, without compensation
- buy it, by providing compensation or purchase on the open market
- save it, by improving infrastructure to reduce transmission and measurement losses
- close dilapidated systems.

The new National Water Initiative will recover water for the environment. This requires sorting out the water market, and then introducing an environmental purchaser to acquire water from willing sellers at the best possible price. This has the advantage of taking water from the lowest value uses, which will be the first to sell, and causes no negative impacts on the higher value uses.

To make the fledging water market work it is necessary to clarify the access rights issue so people know what they are buying and selling. Ideally, we need rights that are consistent across the States so they can be traded across the entire Murray–Darling Basin. At present there are some 20 different water products on the market, all with differing security of supply, and some exist for set periods while others are in perpetuity. We will need market rules that recognise environmental and hydraulic constraints on trade.

There is considerable resistance from some irrigators, especially those producing lower returns for their water. The concern is that water barons will buy up available water and abandon peasant farmers; that water will leave districts and towns; and that infrastructure will be stranded and collapse. There is also a worry that if Government enters the market for the environment the price of water may rise for farmers. These are all valid community concerns that must be addressed by Government.

One approach to recovering the water is a tender approach where we ask irrigators how much of their water allocation they are prepared to sell, at what price. The Governments can then draw the line at the amount of water they can acquire for the money available, and we get it back without the taxpayers being held to ransom.

The emerging urban water agenda

The 2003 COAG National Water Initiative identified the emerging urban challenges but so far little detail is available. There are desirable elements in the urban reform agenda, and I will use Melbourne as a case study, but the experiences are similar in several of our major cities.

Melbourne is a city of around 3.5 million people, projected to increase by a further million people over the next 30 years. Melbourne currently uses 480 GL of water, but with the increased population the demand will increase to 659 GL. The current estimate of the reliable yield from the present catchments is 566 GL. The squeeze will become greater if climate change estimates of a reduction of about 5 per cent in rainfall by 2020 are fulfilled.

Governments have responded by planning for a 15 per cent reduction in water use per head through demand management strategies and an increase in recycling of 20 per cent, both to be achieved by 2010.

Table 6.4 How Melbourne households use their 220 KL/year

Area of water use	% of household use
Garden	35 %
Bathroom	26 %
Toilet	19 %
Laundry	15 %
Kitchen	5 %

Source: Water Resources Strategy Committee for the Melbourne Area, (2002), *21st century Melbourne: a WaterSmart City. Final report.*

Demand management has already led to a significant reduction from some 340 KL a year to the present 220 KL/yr for household water use. This has been achieved by an increase in people living in apartment buildings, dual flush toilets, user-pay pricing, by other water savings and by public education. Further reductions may be more difficult to achieve and will involve changing the way people use water in their gardens.

Table 6.4 shows typical patterns of water use in Melbourne households. It is apparent that over half of the water used in these households is for flushing toilets and watering gardens. This opens up the question as to why potable water is used for such purposes. Urban communities must be encouraged to examine recycling options to replace this potable water with recycled water. There is also scope for water saving, through garden mulching and the use of low-flow showerheads.

Water recycling in Australia has been gaining some momentum over the last decade, with the aim of reducing the environmental impacts of wastewater discharge to streams and coastal areas. There is political pressure to ban ocean outfalls and to make more use of recycled water. This has mainly led to the development of irrigation schemes to use recycled waters for agricultural purposes.

The emerging recycling agenda is, however, to use recycled water to take pressure off the scarce potable water resources by substituting recycled water for potable water. Table 6.5 lists a number of recycling options that are being explored.

We need to develop improved tools to aid decision making in this area. How do we trade-off the impacts on coastal seagrass beds from ocean disposal with the greenhouse gas production from sophisticated wastewater treatment? How do we escape from the sort of economic lunacy where recycled water

Table 6.5 Recycling options for an urban centre

Source of water	Feasibility
Re-use onsite of domestic wastewater	Onsite treatment is possible but requires training and commitment. Not a preferred choice.
Re-use onsite of large building wastewater	Suitable for office or apartments blocks where an on-site plant can be managed by trained staff and used for toilet flushing, water features and air conditioning.
Re-use onsite of industry wastewater	A likely option for large water users if cost structure encourages it.
Sewer mining (use of reclaimed water from sewerage)	Technically feasible for industry re-use or for irrigation of parks and sports grounds. Barriers are cost and consumer acceptability.
Third pipe supply of recycled water (use of an additional pipe to provide recycled water to premises)	Used for agricultural irrigation. Technically feasible for urban greenfield developments. Requires infrastructure at development stage and appropriate pricing.
Collection of roof run-off in tanks	Attractive to users and provides minor benefits to the water supply by taking some pressure off water supply systems.
Use of stormwater	Requires storage in lakes or aquifers. Could be useful for augmenting environmental flows or other uses.

costs $5–7/KL, and yet we sell it for 20c, and are then surprised when usage increases? Why do we sell water from the same dam to irrigators for 20c a KL and to urban residents for 80c, and then try to give them recycled water that costs $5 a KL to produce? When do we install desalination plants and just use seawater that can be produced for probably around $2/KL?

Recycled water is a major challenge to triple bottom line decision-making. People feel strongly about protecting public health. They have great concerns if recycled water is used for personal use but they are relaxed about it in gardens and for toilet flushing. We have some acceptance of third pipe systems to deliver recycled water to households. Balancing the various environmental impacts such as impacts on coastal waters and impacts of energy use on greenhouse gas are not easily costed and so are hard to consider in economic, environmental and social assessments of such decisions.

Conclusions

Australia is entering a period of some water scarcity, and there are pressures to use water more efficiently in both urban and rural Australia. Markets are being used as one mechanism to ensure the water we have is used for the best purposes.

Development in Australia has a long history of profligate water use, particularly from irrigation of marginally fertile lands that was cleared of vegetation for agricultural purposes. The result has been reduction of river flows, increased outflows of fertilisers, herbicides and pesticides, increasing salinity – particularly in lower reaches – and loss of biodiversity. Articulate interest groups seek advantage from water, generally wanting taxpayers to pay the bills and let the users of water avoid the true costs. This situation is continuing with pressures to develop our use of recycled waters where strong social considerations are going to impact on the decisions we make. Making appropriate decisions to do with water will continue to be a challenge for our society. Interest groups will contest the science that is available, especially if they feel it disadvantages them. The tensions between upstream and downstream communities will continue, but considering the entire range of options on a whole catchment basis will give us a foundation on which we can move forward.

Further reading

Anon. (2002). *Australian Catchment, River and Estuary Assessment, 2002.* National Land and Water Resources Audit. Canberra.

Australian Bureau of Statistics. (2004). *Water Account Australia 2000–01.* Canberra.

Cullen, P. (2002). *The Australian Water Experience: Some Ways Forward.* Rosenberg International Forum on Water Policy. Canberra, October 2002.

Cullen, P. (2003). Challenges to the conservation of Australian freshwater biodiversity: An epilogue. *Aquatic Ecosystem Health & Management* **6(1):** 97–101.

Cullen, P. (2003). The common good. In *Uncharted Waters.* (Ed. D. Connell.) Murray-Darling Basin Commission. pp. 49–60. Canberra.

Cullen, P. (2003). Salinity. In *Ecology, An Australian Perspective.* (Eds P. Attiwell & B. Wilson.) pp. 474–88. Oxford University Press, Melbourne.

Murray-Darling Basin Commission. The Living Murray. <www.thelivingmurray.mdbc.gov.au>

Murray-Darling Basin Commission. *The Salinity Audit of the Murray-Darling Basin – A 100 year Perspective.* (1999). Murray-Darling Basin Commission. Canberra.

National Action Plan for Salinity and Water Quality. <http://www.napswq.gov.au>.

National Water Initiative. <www.pmc.gov.au/docs/national_water_initiative>.

Norris, R. *et al.* (2001). Snapshot of River Health. Report to Murray–Darling Basin Commission. CRC for Freshwater Ecology, Canberra.

Water Resources Strategy Committee for the Melbourne Area. (2002). *21st century Melbourne: a WaterSmart City. Final report.* Water Resources Strategy Committee, Melbourne.

7 – Climate change

Graeme Pearman

The best available evidence indicates that global warming is already occurring and that it will continue throughout this century as a consequence of the human production of greenhouse gases. The projected temperature increases are capable of causing massive changes in the distribution of ecosystems across the face of the earth. The challenge is to produce a portfolio of changes to our energy system that will bring about global emission reductions of 70 to 80 per cent over this century. At the same time, we must meet the needs of the two billion who do not currently enjoy access to usable energy, and those of another two billion who soon will be added to the global population. An appropriately literate community is a key prerequisite if Australia is to tackle this agenda for change.

The earth's climate has always varied, and this has had substantive impacts on the evolution and distribution of ecosystems. What is different about this century is that, for the first time, we have been able to observe that a new change has commenced and anticipate, albeit at this stage approximately, that change will continue through the century.

Climate change is, perhaps, the first truly global challenge for sustainability. This chapter is about some of the components of that challenge and questions whether we are well prepared. It is, unapologetically, an idiosyncratic scientist's view and purposefully provocative.

Climate change and the challenges for science

In the late 1980s the United Nations Intergovernmental Panel on Climate Change (IPCC) was established. Its purpose was to periodically assess the contemporaneous science-base of climate change and to report this in a fashion that would assist policy developers around the world to make decisions concerning the importance of the issue, and the nature of intervention required (IPCC 2001a). Working Group I of the Third Assessment Report of the IPCC (IPCC 2001b) gave six key messages concerning climate change:

- observations of global climate give a collective picture of a warming world
- concentrations of 'greenhouse' gases and aerosols (dust) continue to change the atmosphere
- model simulations of the climate system give increased confidence in climate predictions
- observed warming over the last 50 years is likely to have been due mainly to the increase in greenhouse gas concentrations
- human influence will continue to change atmospheric composition (in particular carbon dioxide) throughout the 21st century
- global temperatures (and sea level) are projected to rise by 1.4–5.8°C between 1990 and 2100.

More details of the evidence in support of these findings can be found in Pearman and Hennessy (2003) and with particular reference to recent newspaper accounts of greenhouse 'sceptics' in Manton (2003) and Pearman *et al.* (2003).

While the IPCC Report does consider climate change in relation to Australia, a more recent review (Pittock 2003) summarises the observations of climate change and its consequences specifically for Australia.

The strength of these conclusions has led national and regional governments around the world to establish policy relating to both the adaptations to the anticipated change and to efforts to mitigate against the change through the control of greenhouse gas emissions. Since the release of the IPCC Report, there has also been a substantial shift in the degree of interest and commitment of the private sector in terms of advancing IPCC policies on climate change.

Going beyond the IPCC report, scientists are now engaged with a number of important issues including:

- establishing greater certainty in terms of the scenarios of emissions into the future as these represent at least half of our uncertainty about future warming levels
- translating the global warming projected from our climate models into realistic regional changes that provide the basis for better planning

- accepting that these changes are superimposed on a high degree of natural year-to-year and decadal variability of climate to which many human activities are currently poorly adapted
- maintaining, against an environment of disinterest, the capacity to observe the climate system
- improving understanding of specific processes of the climate system and incorporating this new knowledge into the whole-of-system climate models
- accepting that, inevitably, uncertainties will remain and that the best approach is through integrative risk management where options are chosen on the basis of overt recognition of risk.

At one extreme (perhaps) there are climatologists and geoscientists who understand that in the 12 000 years since the end of the last glacial period and the present, climate warming reflects an increase in global mean temperature of only 5–6°C. In just the last century, however, we have already observed a warming of the planet of 0.6°C, equivalent to around 10 per cent of this change. Indeed, the projections of the IPCC for this next century are for warming of between about 30 per cent and 100 per cent of this glacial–interglacial temperature change in only 100 years. They also know that the glacial–interglacial temperature variations caused massive changes in the distribution of ecosystems (including humans) across the face of the Earth with impacts that continued for thousands of years. (This is well illustrated in the book *The Eternal Frontier* by Tim Flannery (2001), and the reader can enter the wider scientific literature through Overpeck *et al.* (2003) and Pedersen *et al.* (2003).) Climate scientists, therefore, are more likely to consider that allowing the Earth to warm by even 1°C or so over such a short period is 'dangerous' in the sense that we have little idea how natural and human systems can respond to such a relatively large and rapid change.

On the other hand, the majority of the general public who, on a daily and seasonal basis, experience temperature changes of 10°C or more, are more likely to respond that a little bit of warming – in Melbourne, for example – would be good.

We urgently need to examine this issue. It currently divides us, despite the fact that the wording 'dangerous' is incorporated into the objectives of the United Nations Framework Convention on Climate Change (UNFCCC 1994).

A very important component of the future scientific work relates to providing greater guidance as to what 'dangerous' climate change is. This is needed to underpin the level of concern, strengthen commitment to targets for emission reductions and heighten expectations of impacts and adaptive needs.

> Key Challenge: to provide greater guidance as to the definition of 'dangerous' climate change.

Additionally, there are strategic challenges in developing the underpinning science that need to be addressed. These include:

- determining the likelihood that the net uptake of carbon by the global biosphere (which has characterised the global carbon budget of the last two decades) will reverse to become a source of carbon dioxide leading to a greater growth rate of carbon dioxide in the atmosphere
- determining the risk of the thermohaline ocean circulation slowing in the next few decades, which would impact on the rate of oceanic uptake of carbon dioxide and thus the rate of future climate change
- determining the likelihood that warming rates per unit of concentration increase may currently appear slower than they might be in the future, due to the 20th century growth in aerosol emissions (which may have a cooling effect) masking of some of the greenhouse warming.

> Challenge: to provide sound understanding upon which the risk of climate change can be assessed, including the impact of changes in net carbon uptake by the global biosphere, potential slowing of thermohaline ocean circulation, and the masking effects of aerosol emissions.

Challenges for emission reduction and energy systems

The IPCC key message that human influence will continue to change climate throughout the 21st century arose out of a huge research effort related to the global biogeochemical cycling of carbon dioxide between the atmosphere, oceans and terrestrial ecosystems. The carbon dioxide cycle is now reasonably well understood – enough to know that when a molecule of carbon dioxide is released into the atmosphere, it effectively resides there for almost 100 years. The very important message arising out of this finding is that slowing the growth of emissions of carbon dioxide into the atmosphere will not stop the increase of its concentration and therefore will not stop climate change. Scientists can show that reductions of 70 per cent or more in current global emissions are necessary in order to stabilise concentrations.

This is an enormous challenge, as on the one hand, we have probably only a matter of decades, or at the most this century, to get stabilisation under control, yet there are an existing two billion people who currently do not have access to levels of energy that those of us in the developed world use to meet the standards of living we enjoy. Increased access to energy, at least currently, means increased carbon dioxide ('greenhouse') emissions. Further, by the middle of this century, world human population is likely to rise by a further two billion people, each aspiring to the amenities that the utilisation of energy provides, such as heating, cooling, transport and cooking. It is not energy per se that is demanded, rather, it is the amenity that it delivers. Raising energy usage is one option, but the other is increasing the efficiency with which 'amenity' is provided per unit of energy consumed. Indeed, one of the Millennium Development Goals (MDG 2000) relates to poverty alleviation that itself will be in part dependent on the availability of the amenity of energy. These mutual requirements of emission reduction and energy amenity are, on the surface of it, contradictory.

This challenge is so demanding that we can safely conclude that there is no single solution. Rather, there is a need for a portfolio of activities that each delivers part of the answer. At the same time there is a need for maintaining flexibility for different sectors of the global and national communities and through different time scales. The advantage of such an approach is that individual companies, jurisdictions and countries can hone and evolve their respective portfolios (both economic and social) to maximise emission reduction and energy amenity.

Ingredients of a portfolio approach for Australia might include:

- improved end-use efficiency
- higher efficiency combustion technologies
- new automotive technologies
- decentralised power generation
- affordable renewable technologies
- capture and sequestration of carbon dioxide from power plants.

It is encouraging to see the energy sector in Australia paying more attention to the degree to which energy futures will be largely shaped by the demands for emission reduction. They are also recognising that there are opportunities for those who become engaged earlier to build their businesses from these developments. Clearly this is true for the sectors of industry related to renewable energy and gas, but recently also for oil and coal.

The energy futures debate and policy options are currently characterised by a high level of attention to the so-called hydrogen economy, an economy where the main energy source is hydrogen rather than conventional fuels. Often this attention lacks clarity about where the hydrogen will come from and whether the energy to produce hydrogen will be derived from fossil fuels and/or renewable resources. A hydrogen economy is undoubtedly an option for the future, but raises serious questions concerning the likely time of technologies being available at reasonable prices, the potential atmospheric impacts of leaking hydrogen to the atmosphere, the realisation of current energy system investments and the more likely scenario of a pluralistic future for our energy system structure. The hydrogen economy appears to reflect superficial adherence by the media, sections of industry and academia to what is an almost ideological commitment.

This often dismissive and hand-waving approach to a potentially new way of doing things appears to be counter to the challenge for addressing each emerging option in an integrated way. It should entail weighing the trade-offs of wealth generation (and forgone investment returns), social realities and values, a range of environmental issues, and the value for today and the future. A new approach is essential.

> Challenge: to overthrow the traditional approach of sectoral or ideological interests (in industry, the Green movement, in governments) that selects options that suit a particular view and meet a narrow objective and replace it with a much more inclusive assessment of options in which the trade-offs are explicitly declared.

> Challenge: to produce a portfolio of changes to our energy system that bring about emission reductions of 70–80 per cent over this century; meeting the needs of growing population; relieving poverty; and satisfying, as well as possible, the simultaneous goals of wealth generation, social security, environmental protection and intergenerational options.

The energy debate needs to reflect the exciting potential of aspects of the so-called hydrogen economy. Also, through inclusive, integrated systems analysis, it should rigorously address the realities of technological advancements, existing infrastructure, other environmental issues, realistic timing and the value of a pluralistic portfolio of options.

What Australian Federal, State and in some cases, local and city governments are doing reflects the strength of the IPCC report, and the thrust of the International Framework Convention on Climate Change.

This has led to a vigorous debate about whether Australia should sign the Protocol to the Convention (Kyoto Protocol; Kyoto 1997). By and large, this debate has been relatively poorly informed. It is true that, on the one hand, the signing of the Protocol sends a strong signal of commitment by any government to the challenge of emissions reduction. On the other hand, it is also true that contributors to the debate have often failed to recognise the resource, economic and trade imbalances between nations that make it difficult to reach agreements that reflect some degree of equity of responsibility. Further, where economic analyses of options have been applied, they have invariably failed to cope with the admittedly difficult issue of the assessment of costs forgone if climate change is avoided.

It is naïve to think that the commitments and basis of the Protocol are not without political and national influences. Some countries have clearly sought to produce a Protocol that favours their particular economic position in energy trading with respect to others. Indeed, the Protocol is more about the issue of how energy will be traded in the future and national positioning, than about climate change. All nations start from different resource bases (their mixture and amount of resources), different levels of technology and economic development, and perceptions of the impact of climate change on themselves. This leads to highly divergent views on what should or should not be included in such a Protocol.

The fact remains: the global community needs to find solutions to the problem of meeting energy demands and reducing emissions.

> Challenge: through the Protocol or through alternative and perhaps even more proactive and innovative approaches, to set in place the inducements (for markets) and regulations that together lead to significant global emission reduction. These targets need to be for reductions well beyond those within the current Kyoto Protocol targets (which sets a combined target for Europe, for example, of a reduction of around 7% from 1990 levels by the end of this decade). All developed countries need to be looking at reductions of emissions of around 50% by the middle of this century. For Australia, it is likely that the global community will expect substantive reductions of emissions in the future (unlike the increase of 8% allowed in the current agreements).

The UK Energy White paper (Department of Trade and Industry 2003) contains such a target for the United Kingdom and similar targets have been established in Germany and France. The Australian energy white paper (Department of the Prime Minister and Cabinet 2004) does not set a target. Subsequently a target of 50 per cent reduction by the middle of the century has been advocated by the Australian Chief Scientist (Peatling 2004).

Challenges in predicting climate variability

I have worked for many years in atmospheric science, in an institute well equipped to understand and, in recent years, to predict year-to-year climate variability. Throughout, it has been difficult to capture people's attention so that they accept the concept that Australia has a massively variable climate, that we understand some of that variability, and that through both statistical and mechanistic forecasting we could better manage how we live and exploit our resources. Climate knowledge that contributes to foresight would appear to have the potential for multiple benefits for society in a wide range of areas, including water management, insurance, energy demands, health, and improved management of agricultural and natural systems.

> Challenge: to establish better access to predictions of climatological variability through observations and modelling (beyond observational experience) so that the risk of climatic extremes is based on fact rather than personal perception.

Every time we have major drought, there is an outcry (soon forgotten) of 'How do we drought-proof Australia?' There are calls for the use of dubious technologies such as long-range weather forecasting or cloud seeding. (The latter has potential but this is quite limited.) Somehow, it appears that there is acceptance that good and bad years come and go and we are adapted to this. My own position is that actually we are very poorly adapted, for a number of reasons.

First, there is the lack of knowledge of the true characteristics across the whole spectrum of climate variability. It can be argued that at best human population growth and the concomitant expansion in agriculture, water use and economics has been built on about 100 years of hard climatological observation. This means that any natural variability related to physical/dynamic characteristics of the climate system that work on longer time scales is beyond this time period and not incorporated into risk strategies. At worst, expansion in land use through much of the Third World is based on observational records that are much shorter or even non-existent. There the risk of unforeseen climatological extremes ultimately impacting on current developments is potentially high and that could influence the sustainability of these investments in the future. It is too soon to be certain. Modern climate modelling is the only feasible way of addressing the potential of these longer-than-experience probabilities of impacts. Up to now this technology has been poorly utilised.

Second, there is poor utilisation of existing forecasting capabilities, meaning that significant sectors of the community consider weather and climate variability to be a liability they have no control over, rather than a variable they can use when assessing risk and selecting options that minimise impact and maximise economic outcomes. This is not to say that climate variability will necessarily be a major driver of business performance in many cases, but that incorporating such risk into total business risk evaluation promises advantages for those seeking a market edge.

Third, there is incomplete understanding of the interactions between natural climate variability and greenhouse gas-induced warming. A key example in this regard is the uncertainty as to just how much the perceived severity of the current drought has been affected by not just deficit of rainfall – the overly simplistic current measure of drought – but also by the impact of higher temperatures (in a greenhouse-warmed world) on evaporation rates and ultimately on soil moisture availability affecting agriculture and natural ecosystems. For argument's sake, if we knew that half of the productivity loss during this event in the last 12 months was related to that expected from the natural variability of rainfall, and that half was incurred due to increased temperatures related to greenhouse warming, then we would have a quantitative basis for assessing the cost of climate change in dollar terms. It would highlight that the impact of what is perceived to be a 'small' and 'slow' warming may well be felt through extreme events manifesting through non-linear sensitivities to climate. This warming is occurring now and is not something for concern in the distant future.

In fact, it is not so easy to rigorously draw this connection. But a challenge for science is to do so. The challenge for the community is to then incorporate these changes into their planning for future activities. At the moment this is not happening, by and large as a result of a lack of sufficient and convincing understanding.

> Challenge: to better understand the interaction between climate variability and greenhouse warming on extreme events, allowing for improved efficiency of production systems and security of natural ecosystems, while addressing the non-sustainability of current practices and security of cultural and social systems.

Challenges for sustainability science and policy

When each of us makes decisions individually, or through our companies or governments, our choices are generally made in such a way as to balance competing objectives. By and large they might be self-centred decisions, but they tend to recognise that what we do impacts on others or the environment and reflects on both the immediacy of today's outcomes as well as a potential legacy for tomorrow. So today's growing interest in the concept of balancing the triple bottom line is more about doing what we have always done, but doing it overtly and with a degree of rigour that may, in the past, have been restricted to short-term financial decisions or investments. In order to simultaneously maximise wealth, social, environmental and intergenerational equity, we are confronted with the need for new tools that allow for the complexity generated by such demands. We need to make integrated assessments of all of the factors, and, given that we can never have perfect knowledge about the current situation let alone the future, we need these assessments to be in the form of risk assessments related to policy or decision options.

'Sustainability science' is about underpinning these needs. It is about a higher level of integration of natural and social science and indeed the humanities, all under the conditions of rigour and process that characterises science, but directed at delivering options for environmental, social and economic value that overtly balance the competing objectives. We really do not know how best to do this. Improved integrated representations of the whole system need to be underpinned with quality disciplinary understanding and that remains incomplete – continued investment is needed at this level. Continued investment is also required at the even more fundamental level of observations work.

> Challenge: to maintain our ability to observe and understand the systems of the natural and social world, while building higher level, integrative representations of the world for describing policy options.

Finally, even if today we had in place a highly developed science system that balanced observation, process understanding and integration, the value of that system is only delivered if there are workable linkages between the knowledge generation process and policy development. This is the science–policy interface. The reason the interface is so important is because science is unlikely to deliver the best options for policy if it is:

- performed in ignorance of key policy needs, or
- conversely, not sufficiently independent to be courageous in its content, or
- the policy development process itself is segregated into sectoral interest by virtue of the departmental structure of governments or the singular objectives of highly specific company needs.

> Challenge: to establish multiple connections between the knowledge generation process and the policy developers, and to conduct policy decision making in a more whole-of-government, quadruple bottom line framework.

Challenges in building relevant science capacity

Many of the challenges associated with sustainability in general, and the climate-change issues specifically, are not limited by good science and/or technology. However, some are. In any case we need the best knowledge-base possible to underpin policy selection whether we are in the government or private sector. But there are serious questions about what kind of science base we need, and whether current trends in science are conducive to meeting these challenges. The Australian Joint Academies Committee on Sustainability has had a preliminary attempt to highlight some of the components of the science base that is needed in this regard (AAS 2002).

Figure 7.1 presents one way of looking at national scientific and technological capacity with respect to the readiness to deal with issues of global environmental change such as climate change. In the interest of brevity I cannot analyse this framework comprehensively, but make the following points.

An appropriately literate community is a key feature of a nation ready to tackle these issues of change; one that on the one hand is able to understand and participate in the debates about technological options offered by governments or commerce, and on the other hand accepts the value of science in developing such options and thus is supportive of the scientific infrastruc-

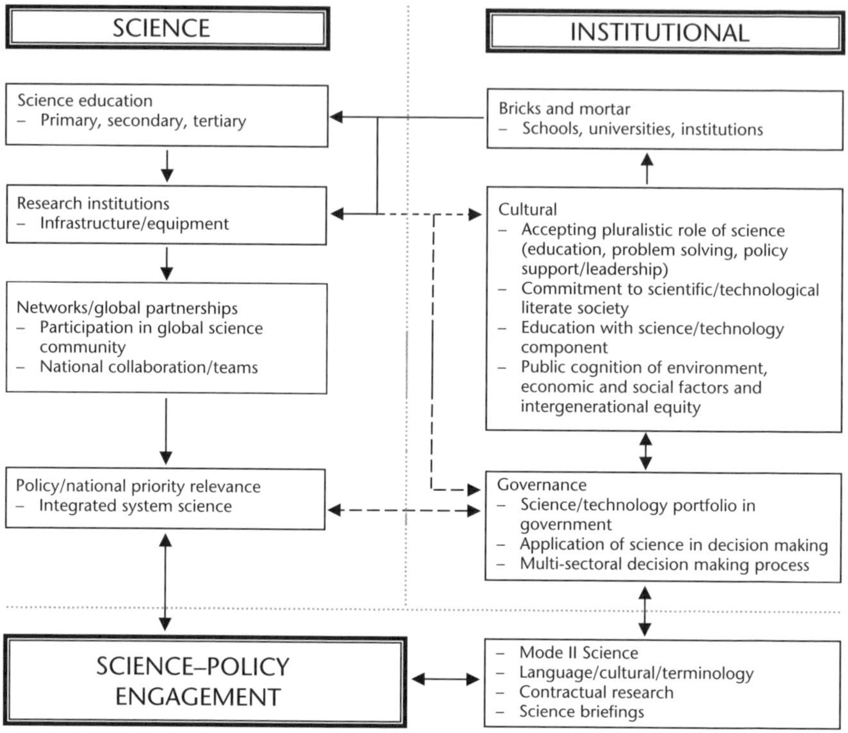

Figure 7.1 Representation of the components of a sound underpinning to Science and its role in the identification of integrated policy options for sustainability.

ture needed for investment in generating knowledge. Thus the scientific community has an educative role in both formal training and informal science literacy that itself is a significant component of setting informed policy and indeed ensuring democratic involvement in that process.

It is argued here that while the current emphasis on the delivery of economic benefit from Science and Technology is a legitimate outcome, there has been a failure on the part of the science community (a weakness symbolised by broken return arrows in Figure 7.1) and of community leaders to recognise this broader role of science in delivering value to the community. That role potentially includes the simultaneous delivery of:

- education
- a window on international science and technology
- policy advice

- solutions to problems
- wealth-generating outcomes, and
- a valuable 'world view' in its own right, as just one other sector of the community.

A failure to capture the full benefits of the investment in the national science effort, through excessive emphasis on one role or the other, undervalues the investment and in turn weakens the commitment of governments and treasuries to make future investment. From this we are all the losers.

Lack of recognition for this pluralistic role of science limits the commitment to science training and institutional diversity. This ultimately affects how attractive or useful science is regarded as a career and vocation – this is not a desirable nor sustainable outcome in itself.

> Challenge: to produce a scientific and technologically literate community and government/private sector administration that recognises the strengths and weakness of Science and its potential role in the transition to sustainability.

But above all, there is the issue of the science–policy interface. There are many ways to have excellent science linked to those who work in the public or private sectors to build policy; in the context of this paper, policy-makers who deliver sustainable outcomes related to the climate issue. Yet most of these mechanisms for the science–policy interface are poorly understood, and often happen by chance or default. There is an opportunity to work more systematically towards the overt recognition and improvement of these linkages.

If the science community enjoyed the best level of commitment and had a well established integrative approach to delivering overtly and rigorously defined policy options, this information would feed into a policy-developing process in government that is itself still highly disciplinary based. Furthermore, government is built upon portfolios that more than often appear to see their best interests, indeed responsibilities, in competition rather than compromise. In the private sector, this is carried to the extreme with highly focused and single-dimensional visions.

> Challenge: to produce multiple avenues for a successful science–policy interface, and governmental/private sector processes that can cope with the objective balancing of social, economic, environmental and equity issues simultaneously through the application of the best Science.

Conclusions

As mentioned above, not all problems concerning the climate change issue and sustainability have their solution in scientific research/knowledge. Indeed, it can be argued that in many cases historical investment, culture, religion and other social factors are 80 per cent or more of the problem in implementing actions that are more sustainable. Scientists do themselves and their communities a disservice to overstate the centrality of science in such debates.

But scientific knowledge, the wisdom that it can enhance and the technologies that it can deliver are important components of the solutions. There is good reason to believe that a new form of integrated systems science can deliver benefits across the sustainability spectrum. The obstacle to science delivering in this way is that many are beginning to look at science as the panacea for wealth generation alone.

Climate change is occurring and will continue through this century. It will be a challenge to manage superimposed climate change in conjunction with existing climate variability, whether it be in agriculture, natural-resource management, water resources, health, commerce or industry.

But one thing is certain: the issue of climate change has changed forever the future of the energy cycle in Australia and the rest of the world. How we are going to meet the growing needs for energy amenity, population growth and, at the same time, tend to existing legacies of unsustainable actions is a huge challenge to all.

Climate change is in a sense a generic test of how we can cope with what is clearly an unsustainable situation. (No one argues that we can accept rapid warming continuing indefinitely.) Climate change is regionally driven with global consequences and is a result of economic imperatives and socially valued goods. It requires a redefinition as to the balance of these outcomes. It includes the dimension that what we do now, almost indefinitely impacts on the options of future generations. What we learn here in our homes, boardrooms and parliaments, about how to manage this situation, will hold value for similar dilemmas to be confronted in our transition to a more sustainable future.

References

Department of Trade and Industry, UK. (2003). Our energy future: Creating a low carbon economy. Energy White Paper: Presented to Parliament by the Secretary of State for

Trade and Industry. <http://www.dti.gov.uk/energy/whitepaper/ ourenergyfuture.pdf>

Department of the Prime Minister and Cabinet. (2004). Securing Australia's Energy Future. <http://www.pmc.gov.au/energy_future>

Flannery, T. (2001). *The Eternal Frontier: An Ecological History of North America and its People.* Text Publishing, Melbourne.

Intergovernmental Panel on Climate Change. (2001a). The Intergovernmental Panel on Climate Change. World Meteorological Organisation and the United Nations Environment Program. <http://www.ipcc.ch/>

Intergovernmental Panel on Climate Change. (2001b). *Climate Change 2001: The Scientific Basis.* (Eds J. Houghton, Y. Ding, D. Griggs, M. Noguer, P. van der Linden & D. Xiaosu.) Contribution of Working Group I to the Third Assessment Report of the Intergovernmental Panel on Climate Change. Cambridge University Press.

Kyoto Protocol. (1997). The Kyoto Protocol to the United Nations Framework Convention on Climate Change. <http://unfccc.int/resource/docs/convkp/ kpeng.html>

Manton, M.J. (2003). Summary of recent reports on climate change science. *Bulletin Australian Meteorological and Oceanographic Soc.* **16:** 130–35.

Millennium Development Goals. (2000). <http://www.developmentgoals.com/ Poverty.htm>

Overpeck, J., Whitlock, D.C. & Huntley, B. (2003). Terrestrial biosphere dynamics in the climate system: Past and future. In *Paleoclimate, Global Change and the Future.* (Eds K.D. Alverson, R.S. Bradley & T.F. Pedersen.) The IGBP Series. pp. 81–103. Springer-Verlag, Berlin.

Pearman, G., Scaife, P. & Walker, B. (2002). A blueprint: The science needed to underpin Australia's transition to sustainability. Australian Academy of Science – Symposium proceedings: Transition to sustainability. Canberra, 3 May 2002. <www.science.org.au/sats2002/blueprint.htm>

Pearman, G.I. & Hennessy, K. (2003). Climate science: what do we know? In: Living with climate change: Proceedings for a national conference on climate change impacts and adaptation. Australian Academy of Science. National Academies Forum, Canberra. pp. 3–31. Available from the Australian Greenhouse Office or the Australian Academy of Science websites or at <http://www.dar.csiro.au/publications/pearman_2003a.pdf>

Pearman, G.I., Hennessy, K., Jones, R. & Maheepala, S. (2003). Climate change and its projected effects on water resources. In: *Water – the Australian dilemma.* Speakers' abstracts. pp. 34–36. Carlton Crest Hotel, Melbourne, Australian Academy of Technological Science and Engineering.

Peatling, S. (2004). Halve gas emissions: top scientist. The *Age* 19 July 2004. <http:// www.theage.com.au/articles/2004/07/18/1090089039419.html?oneclick=true>

Pedersen, T.F., Francois, R, Francois, L., Alverson, K. & McManus, J. (2003). The late quaternary history of biogeochemical cycling of carbon. In *Paleoclimate, Global Change and the Future.* (Eds K.D. Alverson, R.S. Bradley & TF Pedersen.) pp. 63–79. The IGBP Series, Springer-Verlag, Berlin.

Pittock, A.B. (ed). (2003). *Climate Change: An Australian Guide to the Science and Potential Impacts* (electronic publication). Canberra, ACT. Australian Greenhouse Office. viii, 239pp.

UNFCCC. (1994). United Nations Framework Convention on Climate Change. <http://unfccc.int/resource/docs/convkp/conveng.pdf>

8 – Sustainable energy

Andrew Blakers

Solar energy can replace fossil and nuclear fuels over the next 50 years, thus creating a truly sustainable energy supply system. Diversity of solar energy sources is the key to a stable energy supply. Solar thermal, photovoltaics and wind energy are likely to dominate the solar energy mix. Unfortunately, Australian Government energy policy is not nearly so sympathetic to solar energy as is energy policy in Europe and Japan.

Reliable, economical energy supply underlies modern technological society. Now it must be made sustainable.

There are five available energy sources: solar energy (in its various forms), fossil energy, nuclear energy, geothermal energy and tidal energy. Of these, only solar energy can provide really large-scale energy in a sustainable and environmentally acceptable manner. The other energy sources can supplement solar energy to increase the diversity (and hence stability) of energy supply.

Fossil fuels are subject to resource depletion and there is a consensus among climate scientists that the burning of fossil fuels is causing an enhanced greenhouse effect (IPCC 2001; CSIRO 2001). Consequences over the next 50 years could include:

- significant temperature rises (particularly at high latitudes)
- rising sea temperatures and levels (causing flooding, coastal erosion, damage to coral reefs)
- more frequent extreme weather events (such as floods, storms and drought)
- the need to move agricultural activities and infrastructure to different locations
- an expanded range for tropical diseases and disease vectors

- substantial reduction in biodiversity, and
- more severe bushfire seasons.

It is possible that there will be unexpected catastrophes over the next few decades such as loss of the Gulf Stream, due to changes in the circulation of the Atlantic Ocean, and destruction of the Amazon rainforest (see Chapter 7, for more details). Other problems include oil spills, oil-related warfare (for example, the Gulf wars) and pollution from acid rain, particulates and photochemical smog.

Nuclear energy from fission has severe problems relating to waste disposal, reactor accidents, nuclear weapons proliferation and nuclear terrorism. A country that possesses a nuclear energy industry has the raw materials, the technology and the trained people required for the production of nuclear weapons. Several countries have acquired nuclear weapons technology via the route of civilian nuclear energy and so such an energy source carries with it negative implications for nuclear weapons proliferation. Nuclear fusion is still many decades away from commercial utilisation.

Geothermal energy is heat energy that comes from the decay of radioactive elements within the Earth. Heat associated with volcanic regions can be used to generate steam for district heating or to drive steam turbines to produce electricity. Another form of geothermal energy is 'hot dry rocks', which refers to hot masses of slightly radioactive rock buried several kilometres below the surface of the Earth. Cold water forced down to this hot rock becomes steam that can be extracted from boreholes nearby. Geothermal energy is restricted to particular geographical locations. While it is sustainable in the sense that it can be harvested with limited environmental damage, the heat stored in a particular place can certainly be depleted. An Australian company, Geodynamics, is attempting to commercialise this energy source (Geodynamics 2003).

Tidal energy (energy generated from the flow of water during incoming and outgoing tides) can be collected using what amounts to a coastal hydroelectric system. It is sustainable in the sense that it will not run out. However, the coastline is a scarce resource and the collection of large amounts of tidal energy will have a major environmental impact due to flooding of ecosystems by such schemes.

Solar energy, by contrast, is available on a massive scale. Solar energy can eliminate the need for fossil and nuclear fuels over the next 50 years. The collection and conversion methods usually (though not always) entail few environmental or social problems. Unlike fossil fuels, the resource is ubiquitous, which removes questions of monopolisation of energy supply. Solar energy collectors can be spread over a wide geographical area, unlike a nuclear power station, and terrorist strikes on a solar energy collector would not cause massive environmental problems. In addition, a renewable energy system is inherently robust because it comprises thousands of small collectors rather than a small number of large, and potentially vulnerable, generation units.

Solar energy options

Solar energy includes both direct radiation (i.e. harnessing energy from sunlight) and indirect forms of energy (such as biomass, wind, hydro, ocean thermal, ocean currents and wave energy) caused by the effects of the sun on Earth. Most of these energy forms will be part of the energy mix when solar energy becomes the dominant traded-energy form. Some solar energy technologies are more advanced than others. The key to successful mass-utilisation of solar energy is diversity. The solar energy mix will vary from region to region.

Photovoltaics, solar thermal energy (also called solar heat energy) and wind energy are presently the only solar energy technologies that can provide very large quantities of sustainable energy with high (more than 10 per cent) overall efficiency (Blakers 2000). These conversion technologies have small environmental impacts and insignificant military applications. In some countries biomass may also make a substantial contribution to energy supply, despite low conversion efficiency.

Photovoltaics

Photovoltaics is the science of converting sunlight directly into electricity via solar cells without the use of moving parts. It is an elegant but expensive technology. It has found widespread use in niche markets such as consumer electronics, remote area power supplies and satellites. Large numbers of photovoltaic systems are being installed on house roofs in cities. The cost of photovoltaic systems is not a strong function of scale, which means that photovoltaic systems are often the most economical energy source for small applications. Over 90 per cent of the world photovoltaic market is serviced by crystalline silicon solar cells. This dominance is likely to continue for many years.

Electricity production from photovoltaics has been increasing at an average rate of 25 per cent per year over the past decade, which is far in excess of the growth rate in energy consumption. Rapid growth in production is causing steady reductions in cost, which will eventually lead to a true mass-market developing. The current value of annual photovoltaic system sales worldwide is about A$7 billion per year. Australia manufactures about 5 per cent of the world's solar cells.

Solar thermal energy

Solar thermal energy can be harvested in many forms, including high temperature steam, warm air and hot water.

Solar thermal electricity

Most solar thermal electricity technologies use mirrors to concentrate sunlight onto a receiver. The resulting heat is ultimately used to generate steam, which passes through a turbine to produce electricity. Two non-concentrating exceptions are solar chimneys and solar ponds. Concentrator methods are equally applicable to concentrating photovoltaic systems. The usual ways of concentrating sunlight are point focus concentrators (dishes), line focus concentrators (troughs, both reflective and refractive) and central receivers (heliostats and power towers).

Solar thermal electricity is not yet a commercial proposition. The reason for this is that, unlike photovoltaic, there are strong economies of scale. This means that small systems that might be suitable for an individual household are far too expensive. This lack of a niche market, in contrast to photovoltaics, inhibits the development of solar thermal electricity in the short to medium term.

High temperature solar heat

Concentrated solar energy can achieve the same temperatures as fossil and nuclear fuels, either directly (using mirrors) or through the use of chemicals (thermochemicals or bio fuels) created using solar energy. One problem for high temperature solar heat is that heavy industry (e.g. the steel industry) is often located near coalfields, in regions that are relatively poorly endowed with solar energy. Perhaps the next steel mill could be built in the Pilbara region of Western Australia, close to iron ore deposits, rather than on the east coast, close to coal deposits. North-west Australia is one of the sunniest places on earth.

Solar heat can be used to extend fossil fuels. For example, if natural gas (CH_4) is heated using a dish concentrator in the presence of steam (H_2O) then hydrogen and carbon monoxide ($3H_2 + CO$) are produced. The energy content of the hydrogen and carbon monoxide is about 30 per cent greater than that of the methane, and so solar energy has been added to the original methane.

Low temperature solar heat

Low temperature solar thermal technologies use the sun to generate heat rather than electricity. Solar water heaters are perhaps the most well known type of solar thermal technology and are directly competitive with electricity or gas in most parts of the world. Solar concentrator water heaters have been combined with photovoltaic collectors to produce 60–70 per cent efficient hot water and electricity systems. Solar thermal technologies can also be used to heat buildings. Good building design, which allows the use of natural solar heat and light, together with good insulation, minimises the requirement for space heating.

Wind energy

Modern wind generators are very different from old-style water-pumping windmills. For one thing, they are huge. They have 40–70 m high tubular steel towers on a concrete foundation and have three blades, each 20–40 m long. They are computer-controlled and centrally monitored, with many safety features. They have capacities in the multi-megawatt range and will operate reliably for more than 20 years. Wind generator towers occupy very little land per machine. Many wind farms are located on cleared farming land. Farming activities are scarcely affected, and lease fees from the wind generators amount to a second cash crop for the farmer. A large fraction of future wind farms will be located offshore to take advantage of higher wind speeds and to avoid land use conflicts. Provided that a wind farm is not located in an ecologically sensitive area, the only significant environmental impact of wind energy is visual.

Wind energy is likely to generate 10 per cent of the world's electricity by 2020, and is now regarded as a conventional energy source (DWTA 2004; EWEA 2004). The worldwide industry has grown at an average rate of 25 per cent per year over the past decade, and has an annual turnover of around A$15 billion per year.

Biomass energy

Biomass refers to combustible or fermentable biomaterials created via photosynthesis. It can be derived from waste materials such as sugar cane bagasse, garbage, sawdust and sewerage or from energy crops such as trees or canola. Firewood is another common form of biomass energy. When biomass production for energy is combined with other useful purposes the economic viability of biomass energy can be substantially improved. For example, the growth of Eucalyptus Mallee trees in the corridors between fields allows the harvesting of wood by coppicing. The Mallee also serves other purposes including as a windbreak, shade and an ecological corridor. In a country such as Australia, where the population density is low, multiple-purpose energy crops may be viable. In developing countries, where energy use per capita is low, biomass can be a substantial fraction of total energy use.

Unfortunately, the conversion of solar energy into chemical biomass energy has a very low overall efficiency. The conversion of solar energy to chemical energy has an efficiency of less than 0.5 per cent, and conversion to electricity has efficiency below 0.2 per cent (which is two orders of magnitude smaller than from photovoltaics, solar heat or wind energy). When used on a very large scale, biomass competes with food and timber production or with ecosystem preservation for the supply of arable land, water, pesticides and fertiliser. The notion that biomass can be grown at low cost on waste land with small environmental impact is incorrect. The use of low quality agricultural land results in low yields and high costs, whether the crop is for food, timber or for energy. Suggestions that genetic engineering can be used to greatly increase the energy conversion efficiency of biomass are far from current reality.

Other forms of solar energy

Indirect solar energy sources such as waves, ocean thermal gradients, ocean currents and hydro sources are geographically limited. Wave energy technology is still in the development phase, but could be important in the future in regions with large areas of shallow sea and frequent storms. Temperature differences between the deep and surface layers of the oceans can be used to create electrical energy, although there are severe technical problems. Hydro energy is usually associated with large environmental impacts arising from the drowning of river valleys and the alteration of river hydrology. These energy forms could contribute modestly in particular regions to an environmentally responsible energy supply.

Energy issues

In public discussions about energy concerns, a number of issues are raised both in favour and against the widespread use of solar energy. In addition, considerable confusion has arisen in public debate about the roles of hydrogen, fuel cells and carbon sequestration in the sustainable energy economy. This section briefly discusses these issues.

Energy efficiency

Hand in hand with the utilisation of solar energy goes energy efficiency, that is, measures to ensure that the least possible amount of energy is required to complete the task. For example, an energy-efficient building is a building that utilises natural solar light and heat sensibly. Walking rather than driving uses a small amount of solar energy (biomass in the form of food) rather than a larger amount of oil energy. Clothes lines, solar salt production and putting on extra clothing displace electric clothes dryers, fossil fuel fired kiln drying of salt and gas or electric heating respectively. Australia has one of the highest per capita energy consumptions in the world, and major reductions in greenhouse gas emissions are readily achievable from a concerted energy efficiency program.

Energy storage

Energy storage issues are not likely to prove to be major obstacles to mass utilisation of solar energy. However, considerable work will be required on storage once solar energy begins to be a major part of electricity supply. Solar energy is generally less predictable and despatchable than energy supply from fossil fuels. However, fossil fuel generation is not completely reliable. The question to be solved is how to ensure that the probability of failure of energy supply in a solar-dominated energy mix is similarly small to that in a fossil fuel dominated energy mix.

A number of storage options are available, including thermochemical energy storage, pumped hydro, flywheels and compressed air. There are many options for the storage of low temperature solar thermal energy (for water and space heating) that are simple and cheap. Examples include storage in building materials, water, crushed rock and phase-change materials. The latter are materials that melt and freeze at a particular temperature (e.g. 31°C).

Another method of solar energy storage is via chemical reactions driven by concentrated solar heat or light (solar thermochemistry). One example that is being worked on at the Australian National University is ammonia (Lovegrove, Luzzi & Kretz 1999). Ammonia can be disassociated into hydrogen and nitrogen at the focus of a dish solar concentrator system. These gases can be pumped long distances in natural gas pipelines and recombined when convenient to form ammonia and to yield high temperature steam that is suitable for industrial use or electricity generation.

Storage of renewable energy electricity is not a serious issue until the fraction of solar electricity reaches 10–20 per cent of the total. This will not happen (except in a few places) for many years, giving time for improved storage technologies to be developed.

The use of a variety of solar conversion technologies minimises storage requirements. For example, the probability that neither wind energy nor photovoltaic energy will be available at a particular time is lower than the probability that either is not available. Wide geographical dispersal of solar energy collectors, the use of a variety of different solar technologies, the use of energy storage and the judicious use of relatively small quantities of natural gas and bio fuels minimises the probability of a shortfall in electricity supply.

Energy 'payback' time

It is important that the amount of energy produced by a solar energy collector over its lifetime be much larger than the amount of energy used in its manufacture. Some people have erroneously claimed that more energy is used to manufacture solar collectors than is generated by the collector in its lifetime. Fortunately, the time required to recover the energy investment in solar energy equipment is typically one-tenth of the lifetime of the equipment. One exception is photovoltaics, which currently has an energy payback time of about five years compared with an expected system life of 20 to 30 years. However, the energy payback time and cost of photovoltaic systems are closely linked, and both are falling. By the end of the decade the energy payback time of photovoltaic systems is expected to be less than two years.

Hydrogen

A great deal has been written about the 'hydrogen economy', the idea that hydrogen as a fuel could stand in the place currently occupied by oil and gas. Hydrogen is a gas with a very low boiling point (-253°C). There are several technical problems associated with hydrogen use, including the difficulty of storage. Hydrogen storage could be by way of compression or liquefaction, or by reversible sorption processes or chemical reactions. Storage for use in cars using methods that could compete economically with petrol or compressed natural gas is a challenging proposition, but is receiving considerable attention.

By far the largest problem is the question of the source of the hydrogen. Hydrogen is currently derived by reacting water with fossil fuels at high temperatures. This produces carbon dioxide and is no more sustainable than any other fossil fuel burning technology. Electricity can be used to split water into hydrogen and oxygen (electrolysis). In principle, the electricity can be from a sustainable solar energy source. Unfortunately, solar electricity is relatively expensive. Electrolysis of water to produce hydrogen entails substantial conversion losses, and conversion of the energy in hydrogen to any form of energy other than heat entails further substantial losses. In most cases it is cheaper and more efficient to use the renewable electricity directly. The direct splitting of water under sunlight (e.g. by using titanium dioxide) has formidable technical obstacles, relating to corrosion and very low conversion efficiency, which are far from resolution.

Fuel cells

Fuel cells convert gaseous fuels (e.g. hydrogen or natural gas) to electricity without combustion. Fuel cells are being explored as replacements for fossil fuel generators of electricity as they have the potential to be considerably more efficient than conventional combustion, particularly in small systems. Substantial technical obstacles still remain. Fuel cells are sometimes claimed to be a renewable energy enabling technology because they could convert hydrogen energy (produced by electrolysis using solar energy) to electricity at relatively high efficiencies. However, there are many conversion losses in this sequence.

Fuel cells may have important applications in saving energy. For example, the use of fuel cells in houses to produce electricity from natural gas, with space and water heating as a by-product, could considerably reduce total greenhouse gas emissions from houses, particularly in cold climates.

Solar energy for transport

Energy for transport is one of the most difficult markets for solar-derived energy to penetrate. Liquid fuels from biomass (e.g. ethanol) can power vehicles, but only at substantial environmental cost if used on a large scale. The private car in cities can be largely replaced by public transport, which is much more energy efficient than a car and can be powered by renewable electricity. Freight can be shifted to electrically powered trains. Lightweight electric cars are more efficient than current automobiles for city use. It is an open question as to whether significant private motor vehicle ownership can be afforded in an environmentally constrained world.

Carbon sequestration

Carbon sequestration is a means to stop the carbon dioxide created by burning fossil fuels from entering the atmosphere. It has been touted as a means by which Australia's large coal resources could be utilised for many more decades without adding substantially to the enhanced greenhouse effect. One method of sequestration is to separate the carbon dioxide from the products of combustion, compress it and pump it into saline aquifers. Another method is to capture the carbon dioxide using rapidly growing genetically modified plants. The plants use sunshine, carbon dioxide and water to make biomass. This biomass can either be buried in the mine from which the coal came, or dried and burnt in a power station to produce electricity. This method is, of course, a solar energy technology.

Current estimates for the cost that 'zero emission coal' electricity might reach over the next decade or two are comparable with estimates for various forms of solar electricity. Carbon sequestration may assist the transition to a greenhouse-neutral energy economy in the period to 2050. The fossil fuel industry is devoting considerable resources and political capital to the development of methods of storing carbon dioxide from power stations underground or in the ocean.

Sustainable energy in Australia

Solar energy collection technology is immature compared with fossil fuel technology, and solar electricity is relatively expensive compared with energy from fossil fuels. As the solar industry grows in scale and as the technology matures the gap in cost is steadily decreasing. Fossil fuel users currently have a free licence to pollute by emitting carbon dioxide. When the cost of eliminating greenhouse gas emissions is added in, for example by requiring full geosequestration of the CO_2, then solar electricity production is likely to be fully competitive with electricity production from fossil fuels.

Australia is a great place to harvest sustainable energy on a mass scale. We have excellent and diverse solar energy resources by world standards. Australia has a low population density, is one of the sunniest countries in the world and has vast wind resources, particularly along the southern coastline.

Several solar-based renewable energy industries (e.g. solar hot water, photovoltaics and wind energy) are likely to be A$100 billion per year (worldwide) industries by 2015, and Australia has a foothold in the renewable energy industry. BP Solar and Rheem/Solahart have major manufacturing facilities in Australia for photovoltaics and solar water heaters respectively. Several wind energy companies are considering the establishment of assembly plants in Australia for wind generators. Origin Energy is commercialising the Sliver® solar cell technology developed at the Australian National University and Solar Systems is commercialising its dish-concentrator photovoltaic technology. Australia has two large world-class renewable energy research groups: the Centre for Sustainable Energy Systems (CSES) at the Australian National University (CSES 2004) and the Key Centre for Photovoltaic Engineering (KCPVE) at the University of NSW (KCPVE 2004). In addition, a number of small groups and companies are ready to respond quickly to improved funding for renewable energy.

Australia has large coal, oil and gas deposits and current Australian energy policy is strongly influenced by the coal industry. The Australian Government has committed large resources to strategic fossil fuel R&D through a variety of mechanisms, including Cooperative Research Centres, CSIRO divisions, Geoscience Australia and the Rio Tinto Foundation. Carbon sequestration is a particular focus. Far more government funding is committed to support for the fossil fuel industry than to support for the renewable energy industry. It is a risky strategy for the Government to direct most of its efforts to reduce energy-related greenhouse gas emissions towards carbon sequestration.

The prospects for substantial Australian participation in the international renewable energy industry will depend strongly upon the energy policy of the Australian Government. If the current focus on fossil fuels is maintained then Australia will be a bit-player. On the other hand, the natural advantages of Australia for solar energy utilisation, coupled with our strong technical base, would allow the Australian renewable energy industry to respond vigorously to favourable policy settings.

In order to rapidly grow the Australian renewable energy industry, the Government needs to provide internationally competitive support for both R&D and markets. In addition, a clear commitment by Government to 50–80 per cent cuts in greenhouse gas emissions by 2050, with interim milestones, would galvanise industry.

If we Australians choose to follow the European example by adopting a low-greenhouse-emission energy policy then it will be neither particularly difficult nor particularly expensive to implement over five decades. This requires leadership and clear and consistent policy. Science and industry will deliver the required technology.

Conclusion

Wind energy, solar thermal and photovoltaics are the only truly large-scale sustainable electricity generation technologies available. They are each likely to be $100 billion/year industries by 2015. These technologies are relatively free of adverse environmental impacts, and will come to dominate the traded energy market over the next 50 years.

It is likely that international concern over the enhanced greenhouse effect will continue to increase. The consequence of this concern will be ever-increasing support for solar energy around the world. It is to be hoped that Australian Government policies will be such as to place Australian companies in the forefront of this rapidly growing industry.

At present Australian energy policy is being driven by short-term considerations in favour of the fossil fuel industry, to the detriment of the renewable energy industry. Dedicated and strategically directed funding of solar energy R&D and research and professional training, together with reliable long-term commercialisation support for Australian-based manufacturing, is required if Australia is to become a major player in this vast new industry.

Further reading

Blakers, A. (2000). Solar and wind electricity in Australia. *Australian Journal of Environmental Management* **7**: 223–36. <http://solar.anu.edu.au>

CSES. (2004). Centre for Sustainable Energy Systems, Australian National University. <http://solar.anu.edu.au>
CSIRO. (2001). Climate Change Projections for Australia. <http://www.dar.csiro.au/publications/projections2001.pdf>
DWTA. (2004). Danish Wind Turbine Manufacturers Association. <http://www.windpower.dk>
EWEA. (2004). European Wind Energy Association. <http://www.ewea.org>
Geodynamics. (2003). Geodynamics Ltd. Milton, Australia. <http://www.geodynamics.com.au/IRM/content>
IPCC. Intergovernmental Panel on Climate Change: <http://www.ipcc.ch>
KCPVE. (2004). Key Centre for Photovoltaic Engineering, University of NSW. <http://www.pv.unsw.edu.au>
Lovegrove, K, Luzzi, A, Kreetz, H. (1999). A solar driven ammonia based thermochemical energy storage system. *Solar Energy* **67(4–6):** 309–16.

9 – Urban design and transport

Peter Newman

Australian cities are under pressure as a result both of global climate change and their heavy dependency on oil. The goal of sustainable urban development is to reduce the ecological footprint of the population while simultaneously improving the quality of urban life. Our cities are currently among the world's worst consumers of transport energy. Car dependency is inherent in policy thinking across the nation but there are growing efforts to change this. We need to create viable population centres that will support enhanced public transport systems. This means increasing the population density of parts of the cities. New funding and policy are needed to move cities towards sustainability. Australian building approval processes need to specify reduced energy and water use and ways need to be found to involve community members in ownership of the transition to sustainability.

Sustainability arises from a new sense of limits, that the globe's ecosystems, resources such as oil and the climate system, are being overstretched. Most of this impact is coming from the world's cities. Sustainability approaches such limits with a positive mindset, believing that opportunities can be created to achieve economic and community gain while reducing our ecological and resource impacts. In this way sustainability in settlements can be seen as an opportunity to reduce our ecological footprint whilst simultaneously improving the quality of urban life.

There are three main limits that have emerged in Australian settlements in recent years:

- **Water**, where nearly every major city in the country has had to institutionalise processes to reduce water consumption, and some scenarios of serious long-term supply difficulties due to greenhouse-induced climate change have emerged for our cities

- **Oil,** where the peak in global oil production is emerging at the same time as indigenous supplies of oil are reducing and the prospects of significant dependence on Middle Eastern oil has created a major vulnerability for transport systems in our cities
- **Travel time,** where urban sprawl has reached such a distance from major centres and services that the cessation of fringe development, better transport options and new urban design solutions have become a deeply felt need in Australian cities.

These limits are linked to others such as the need for less greenhouse intensive urban lifestyles and an overall dominance by cars in our cities that have many social, environmental and economic implications. These limits will be discussed below as the basis for change towards sustainability in Australian cities.

Together these limits suggest two key challenges for Australian settlements:

Challenge 1:	The need for more ecological building innovations and more ecologically responsible urban lifestyles (to save water and energy especially)
Challenge 2:	The need for less car-dependent urban design and less car-oriented lifestyles (to save oil and travel time especially).

Furthermore, two other overall challenges underlie these first two:

Challenge 3:	The need for a new government funding process for sustainable cities in Australia
Challenge 4:	The need for true participation on the long-term sustainability of our cities.

Some key policy conclusions for Australian cities to be more sustainable will then be deduced.

Challenge 1: Encouraging ecological building innovations in cities

Ecological innovation is a key component of sustainability in our cities that needs close attention by governments at all levels. Saving energy and saving water have taken on a new political immediacy in recent years as limits to electricity supply have become necessary for financial and environmental reasons, and water supply limits have loomed large as droughts have extended and deepened across southern Australia. For 30 years we have known how to build in a more energy and water-efficient way but the systems of land development and building have not considered it necessary to regulate sufficiently in this direction. Changes in July 2003 introduced in the 'Building Code of Australia' have more substantial energy efficiency requirements for building. Since then various State Governments have been examining how to extend this. In particular the water issues in Australian cities have become far more dominant and hence there has been a clear political basis for introducing tougher rules on how to build in water efficiency. Both NSW and Victoria have introduced innovations in standards for water and energy use in buildings.

The next stage is to create an Australia-wide statutory development control system that can be applied to all new development and all redevelopment. It should also involve all the features of ecological innovation: energy, water, site ecology, building materials, waste management from buildings and so on. One such system called BASIX is being developed in NSW (www.iplan.nsw.gov.au/basix). A 'sustainability scorecard' for all new urban development that is based on outcomes, BASIX encourages innovation to achieve these goals rather than being a too-prescriptive set of rules on how to do it. The partnerships being used to develop and trial the system appear to be a model for how sustainability can be developed as a progressive exercise in public policy.

In July 2004 the NSW government introduced BASIX certificates for energy and water outcomes and it will progressively move towards the other areas of ecological innovation. All new and renovated developments in Sydney are now required to use 40 per cent less water and 25 per cent less energy than average Sydney households of the same type. There is a need for a national approach to this issue so that developers and communities can be assured of a common set of goals without arbitrary rules.

There are many good sustainability rationales for promoting this kind of innovation in ecological building. However, it is important to see that such building is a global market and if Australia can do it well then the new technologies, building materials (e.g. eco-cement and other recycled materials) and support systems, are all good exports in a world hungry for sustainability.

Such innovation in ecological building will only work if people also actually use such buildings with an ecologically sensitive mind-set. If householders abuse their new energy efficient homes they can still have excessive energy consumption. Though a new suburb may be built with water sensitive urban design, low water appliances and low water gardens, householders can still waste water. People will need educational assistance to embed a conserver lifestyle. Such assistance can be provided and there is evidence that with targeted programs such changes can be made for long-term gains (McKenzie-Mohr & Smith 1999).

Challenge 2: Reducing car dependence

People in cities have a fixed travel time budget of around one hour per day. This Marchetti Constant has been found to apply in all cities and throughout urban history (Marchetti 1994; Newman & Kenworthy 1999). People seem to have a biological need for restorative or reflective time that is not work or domestic activity and in history this was based around walking between activities. It helps us to understand why old walking cities were only 5 to 8 km across as average journeys needed to be within this time band. It explains why transit cities built around trams and trains spread out 20 to 30 km and car-based cities spread 50 km. It also shows why we are reaching the limits of urban sprawl as cities go beyond this distance. And why problems of obesity and local identity are not assisted by travel time budgets being totally absorbed by time in a car rather than time given to walking.

Time limits are emerging in cities as we see growing road rage associated with congestion, cities turning back in from the sprawl that has characterised them in the past, and the rapid growth of Transit Oriented Development (housing and commercial development within walking distance of a railway station) as a preferred location for people to live and work so that time can be saved in the city (Newman 2003).

Car dependence can be understood in terms of this travel time issue. Cities which have built only car-dependent suburbs generally have slow public transport and unprioritised walk/bike options for local transport. This means that they cannot provide an answer to the travel time budget for most people without the use of a car. To overcome car dependence, travel options are required where public transport is faster than other traffic down all main corridors, and local transport options must favour walking and cycling for short journeys. Land use to support this means more intensive residential and employment activity in centres along the main corridors.

These policy priorities are based on a number of ideas that have been argued out over many years in Australia and are now reasonably accepted by the general public (see for example Laird *et al.* 2001, and the Australian State of the Environment Reports for 1996 and 2001, as well as Warren Centre, 2003). These ideas are that:

- cities are shaped by transport infrastructure
- car dependence has been the dominant Australian urban paradigm for 50 years and so our cities are now excessively sprawling with inadequate infrastructure for alternative, more sustainable transport options
- most Australian cities are trying to avoid car dependence in their planning systems, e.g. Melbourne's 2030 Strategy, Brisbane's SEQ 2021, NSW's Strategic Plan and Perth's Network City
- but most cities have difficulty resolving density issues and funding of alternatives to thc car.

Reducing car dependence requires a combination of better infrastructure for sustainable transport options – trains, buses, bikes and walking – and encouraging land use that fits this infrastructure. In particular, urban centres are needed that are more than just shopping centres. There are plenty of underused rail facilities in Australian cities where it makes sense to build housing and jobs nearby so that people can use the services already in place. There are also new suburbs that can be built around true urban centres that encompass residents and commercial activity, and provide new residents with sufficient local shops, education, health and other services so that they can walk, bike, bus or take a short car ride rather than travel longer distances for these services. Such centres are best developed around a rail node so that users can still reach the rest of the city for other urban services (e.g. universities, major health facilities and specialised services) without needing a car.

This approach to reducing car dependence means that people can still own and use cars when they truly need them but they are not so totally dependent that they are chained to a car. Australian suburbs are full of families with far too much of their total income going on transport (in some cases this is rising to 40 per cent) due to a deep car dependence that has now become an entrenched culture of unsustainable activity. The data on cities we have collected over the years is very clear on car dependence: it is not good for the economy of cities, the environment of cities or the community of cities.

Economically, Australian cities are using around 13 per cent of their city wealth on transport. This compares with European cities at 8 per cent and wealthy Asian cities like Singapore, Tokyo and Hong Kong with just 5 per cent (Newman & Kenworthy 1999); these cities also have strong and well-patronised public transport systems. US cities that are similarly car dependent like Australian cities are spending 12 per cent of their city wealth just getting around. In other words the cities with the best public transport and least car dependence are working better economically.

Nor are the community benefits of car dependence in any obvious way an offset to these economic costs. A recent study of children being driven to school in London shows that they have much reduced risk-taking ability and community development characteristics compared to children who walk to school. Hillman (1999) has shown that this is due to the lack of sense of place and the development of trust in neighbours and strangers, which is directly related to the way that cars cut us off from local community. Tranter (1993) has documented in Canberra the decline of school children's travel habits into car dependence and predicted many social consequences.

Environmentally Australian cities are amongst the world's heaviest consumers of transport energy, producing smog emissions and greenhouse gases as a consequence and sprawling over considerable areas of land. US cities like Houston and Phoenix use over 60 GJ of transport energy per person, Atlanta uses over 100 GJ, and New York's use is 46 GJ per person. While major Australian cities are generally better than those of the US, the worst Australian cities, Canberra and Perth, use 40 GJ per person and the figure for Sydney is 30 GJ. There is considerable variation evident even in the car dependent cities. European cities, however, use much less transport energy – between 13 and 25 GJ (around half that of the car-dependent US and Australian cities); and the wealthy Asian cities use between 3 and 8 GJ per person (Newman & Kenworthy 1999).

These patterns of car dependence across the world's cities are mirrored inside Australian cities in terms of access to public transport and environments where it is easy to walk or bike to destinations. In Melbourne, for example, there is a major difference in the travel habits of people in the inner/core compared to the outer/fringe which is of the same order as between say US/Australian cities and European cities overall. People in the inner/core take on average 2.12 car trips per day and 2.62 walk/bike trips but those in the outer/fringe suburbs take 3.92 car trips per day and just 0.81

walk/bike trips. As well the inner/core has three times the level of public transport use with 0.66 trips per person per day compared to 0.33 trips in the outer/fringe. In Sydney 17 per cent of people in the outer suburbs use public transport for work but in the inner suburbs it is 33 per cent.

To make it worse, the people who are most car dependent are the least able to afford it. So the question boils down to how we can begin to reduce car dependence in Australian cities, especially in the outer areas, in ways that do not make inequalities worse.

The models of how Australian cities need to be planned are being set out in all the planning systems mentioned above. They are showing the importance of new, more viable subcentres, especially in outer areas, linked by quality public transport routes (both into the city centre and around it) that are faster than the car traffic in those corridors. They also reflect a clear policy of restraining growth at the fringe and therefore concentrating development into these centres.

Overcoming car dependence then comes down to two major issues: (1) how to increase the viability of alternative transport modes in corridors and in local areas, and (2) how to create viable centres that will enable good public transport to be provided and to encourage sufficient services locally in such centres.

The first issue is reasonably well known and is just a question of funding priorities (see below). The second is less well known. From data collected globally and locally there is evidence of a minimum level of density of population and jobs (development intensity) that needs to be reached before car dependence can be adequately addressed. This minimum development intensity level is around 35 people and jobs per ha (Newman 2004). Once the level of activity reaches this kind of intensity then the level of car use drops dramatically and alternative sustainable transport modes begin to be genuine options. Similar patterns are found in the world's cities where a 35 per ha minimum seems to be associated with car dependence in general.

What this means in urban design terms is that the number of people and jobs in centres needs to be planned to try and reach this level. If a small centre were to be built around a rail station then a viable number of people and jobs within 1 km radius of the centre would be around 10 000. If a major regional centre was to be built it would need around 100 000 people and jobs within a 3 km radius for it to be a viable centre. If centres don't reach these numbers then they need to bring people from further out and the time/space limitations of transport would mean the centres then became car dependent as with outer area shopping centres. The levels of density are simply related to the levels of amenity (access to services) that can be viably provided in a centre. Most recent Australian suburbs have been built without the levels of amenity present in older, less car-dependent suburbs (see Newman *et al.* 2003).

The proposed 'Transit Cities' of the Melbourne 2030 Metropolitan Strategy have around this level of activity intensity and would therefore help considerably to reduce the car dependence of the city (Victorian Government 2003). The 13 transit cities in the plan will bring urban amenity similar to that provided in most inner areas. Melbourne, like many other Australian cities, however, finds itself with local resistance to this level of activity from those in the areas where these transit cities are being planned and thus processes to examine the sustainability of the plans must be able to bring local people with them if such transformative vision is to be achieved.

These new centres in the suburbs as well as the present centres in the inner areas ought to also have a clear commitment to providing social housing (see Newman 2002). Otherwise, the growing inequities in access to amenity in Australian cities could be made even worse.

In addition to policies of improving public transport infrastructure and building better centres, policy to contain urban sprawl by setting urban growth boundaries is an essential feature of transforming a car-dependent city into one where sustainable transport has some kind of possibility of succeeding. Melbourne and Adelaide have recently brought in urban growth boundaries and Perth and Sydney are pursuing the idea. A city like Perth has always had a statutory boundary for development but this has been far beyond the growth front and has never acted as a stimulus for more redevelopment around centres as suggested above.

In parallel to the need for reduced car dependence in Australian urban design there needs to be an emphasis on reducing car-dependent lifestyles. Even in the best-designed neighbourhoods people can use a four-wheel drive to go 200 metres for a litre of milk. There is growing evidence that programs like TravelSmart (www.travelsmart.gov.au) and other voluntary behaviour change methods can create the changes required to wean people off their cars (James 1998; Ampt 2003). Such programs need funding support.

Challenge 3: New funding processes for sustainable cities

Urban design innovations for sustainability are simultaneously an issue for transport infrastructure, for urban planning and for urban behaviour. The transformation of centres suggested above will not happen unless it is simultaneously part of a re-building and improvement of sustainable transport infrastructure. The new Subiaco Centre in Perth, which features dense mixed-use development around a railway station, is being seen as a model of how to build a 'Transit City' centre (Newman 2001). However, this would never have been able to attract financial investment by the private sector unless it had been associated with the upgraded rail service from Perth to Fremantle and the Federal ALP Government's 'Better Cities' plan which helped to fund the redevelopment of the train station precinct. The resulting area has been laced with bikeways and footpaths, and with dense apartments selling off the plan around the station there has been a 100 per cent increase in patronage of the station. Such stories are common when 'Better Cities' projects are examined across Australia.

Such a model of public–private partnership has not been evident in Australian cities since the demise of the 'Better Cities' program and the complete abrogation of any Federal Government leadership for sustainability in our cities. Yet it is also necessary to say that at no stage in the past 50 years of Federal Government has transport funding been focused on anything other than promoting car dependence. 'Better Cities' was in fact a Housing initiative. The data on funding for transport from the Federal Government between 1975 and 1998 was:

- $42b Roads
- $1.2b Rail
- $1.3b Urban Public Transport.

The case for a better balance in funding towards more sustainable modes has many rationales. Laird *et al.* (2001) show that there is a 'road deficit' in Australia due to the excessive bias towards cars and trucks (Table 9.1).

The other rationale for a change in funding is the issue of oil vulnerability. The global oil production peak is either near or has already passed (Campbell 2003). The end of the age of oil is upon us. For Australian cities and transport systems there is an urgent need to move towards a rapid transition into greater use of gas and renewable fuels. However, it will not be sustainable or possible without significant reductions in car dependence as part of the transition. This is the most important policy transition to make that can assist the world to be less dominated by terrorism as oil politics lies at the heart of many of these global disputes (Heinberg 2003).

Table 9.1 Income and expenditure due to Australian transport funding, with resulting 'deficit'

Expenditure
• $7.0b Road construction and maintenance
• $15b Road crashes (health and personal loss, and time loss)
• $3.0b Other health (noise and air pollution)
• $3.3b Refunds and subsidies (tax concessions)
Total expenditure = $28.3 billion
Income
• $8.5b Fuel excise
• $3.8b Registration fees
• $8.0b Insurance
Total income = $20.3 billion
Net Road Deficit = $8.0 billion

Note: the year of these figures is late 1990s; the deficit will be worsened by decisions in subsequent years to reduce taxes on diesel.

Source: Reproduced from Laird *et al.* 2001 with permission of the University of New South Wales Press.

There is a clear need for a better system that can target the innovative changes that are needed for more sustainable cities and sustainable transport in Australian cities, as well as behaviour change programs discussed above. The new Federal Government transport funding system has finally incorporated rail freight into its system. However, there is a complete vacuum when it comes to cities and projects like urban rail and the kind of innovations needed for the renewable transition. Cities in Australia in the knowledge economy are the basis of our participation in the global economy; they will not compete well unless they are more sustainable with less overall car dependence. A new funding system is required to enable this transition to occur. If transport and housing funding were redirected into a new 'Sustainable Cities' program (Newman 2002) there would be sufficient funds to provide innovation and demonstration sustainability projects.

Challenge 4: Encouraging true public participation in planning

It is not possible to make the kind of transitions outlined above unless the public in our cities is part of the process. Planning is inherently democratic in its processes but has lost a lot of common involvement in the issues of transport and sustainability. Experts in bureaucracies have become too dominant on the major planning issues impacting on our lives.

Table 9.2 Vision priorities from the Dialogue for the City process in Perth, 2003

Strong local communities (city of villages)
Clean, green city
Urban growth boundary
Connected, multi-centred city
Reduced car dependence – better public transport, especially more rail, better local bike/walk and integrated transport/land use
Housing diversity (more options)
Access to city services for all

Source: Dialogue for the City process, Perth 2003 (www.dpi.wa.gov.au/dialogue).

Whenever the public are asked their views they tend to be very keen on the idea of sustainability, particularly new public transport options. For example, a government agency survey in Perth showed that 78 per cent of people wanted transport investment to be for sustainable transport and 87 per cent said that it could come out of road funds.

In all Australian cities there are the beginnings of deliberative democracy approaches to planning. In local areas tools such as Design Charettes (intensive integrated design workshops) are now common to help generate local visions for thorny planning issues. And there are new techniques for strategic planning such as Citizen Juries, Delphi techniques and Backcasting from Scenarios. A deliberative democracy approach to planning the city was held in Perth, called 'Dialogue for the City', involving some 1200 people in an exercise to set out what they wanted to see prioritised in the city over the next 30 years (www.dpi.wa.gov.au/dialogue). As Table 9.2 shows most of the visions highlighted in this chapter were the public's preferred future.

There can be little lost from greater public participation in the transition of Australian cities towards sustainability. Indeed a case can be made that the public is likely to be more forthright about the necessary changes than the bureaucracies that have grown up based around car dependence and inadequate consideration of ecological innovation. The issues to do with the intensity of development seem likely to be the most contentious, but even here the public can see why it is needed if it is explained and if true 'common good' outcomes are seen to be associated with the results. Engaging the public in issues such as where people live, the way they move around and the nature of their dwellings is essential if sustainability outcomes are to be achieved.

Conclusions and recommendations

Australian cities are too car dependent and show few of the necessary ecological innovations of sustainability. They lack the necessary government processes, especially at Federal level, to bring about a transition to sustainability in cities and they are only just beginning to develop the true participatory processes needed for cities to grasp a more sustainable future. Set out below are recommended policies for achieving more sustainable cities in Australia.

- Reduce car dependence through the development of new 'Transit City' centres in the suburbs and upgraded sustainable transport infrastructure. Such concepts are appearing in State-based plans but have insufficient funding and inadequate public support processes.
- Fund innovative 'Sustainable Cities' demonstration projects with re-directed housing and transport funds. This should be led by the Federal Government in partnership with States and local government, and with major private sector involvement. These innovative projects would ideally involve transport infrastructure integrated with land development and behaviour change programs.
- Encourage Australian building approval processes that contain 'stretch' sustainability outcomes. Like the NSW BASIX system these should enable the average house (and commercial building) to use 50 per cent less extrinsically produced energy and 50 per cent less water as well as other sustainability outcomes to do with waste, site ecology, transport and affordability. Renovation of the present building stock should also be included.
- Provide sustainable city innovation R&D funds. Australian building and transport innovations that can contribute global solutions towards sustainability need to be promoted by the Federal Government in partnership with the private sector.
- Develop deliberative democracy-based visions and policy for sustainable cities. Transparent and engaging processes for innovative planning in our cities and transport systems are essential for settlements to reach their potential for innovation in sustainability.

Transport and urban design shape the way we live in cities. Australian city dwellers have unacceptably high ecological footprints. As the majority of the world now lives in cities it is feasible for urban Australia to make a real global contribution to sustainability by demonstrating how we can change.

References

Ampt, E. (2003). *Voluntary Household Travel Behaviour Change: Theory and practice.* Steer Davies Gleave, Sydney.

Australian State of the Environment Committee (2001). *Australia State of the Environment 2001.* Independent Report to the Commonwealth Minister for the Environment and Heritage, CSIRO Publishing on behalf of the Department of the Environment and Heritage, Canberra.

Campbell, C.J. (2003). *The Essence of Oil and Gas Depletion.* Multi Science Publishing, Brentwood, Essex.

Heinberg, R. (2003). *The Party's Over: Oil, War and Fate of Industrial Societies.* New Society Publishers, Gabriola Island, Canada.

Hillman, M. (1999). *The Impact of Transport Policy on Children's Development.* Policy Studies Institute, London.

James, B. (1998). Changing travel behaviour through individualised marketing. In: *The Benefits of Green Transport Plans.* Department for Transport, UK Government, London.

Laird, P., Newman, P., Kenworthy, J. & Bachels, M. (2001). *Back on Track: Rethinking Australian and New Zealand transport policy.* University of NSW Press, Sydney.

Marchetti, C. (1994). Anthropological invariants in travel behaviour. *Technical Forecasting and Social Change* **47(1):** 75–8.

McKenzie-Mohr, D. & Smith, W. (1999). *Fostering Sustainable Behaviour: An introduction to community-based social marketing.* New Society Publishers, Gabriola Island, British Colombia.

Newman, P.W.G. & Kenworthy, J.R. (1999). *Sustainability and Cities: Overcoming Automobile Dependence.* Island Press, Washington, DC.

Newman, P.W.G. (2001). Railways and Re-urbanisation in Perth. In: *Case Studies in Planning Success.* (Eds J. Williams & R. Stimson.) Elsevier, New York.

Newman, P.W.G. (2002). Housing and Sustainability: More than a roof overhead. Ninth Annual Barnett Lecture, Ecumenical Community Housing, Uniting Care, Hanover, Anglicare and AHURI, Melbourne.

Newman, P.W.G. (2003). Global Cities, Transport, Energy and the Future: Will ecosocialisation reverse the historic trends? In: *Making Urban Transport Sustainable.* (Eds N.P. Low & B.J. Gleeson.) Palgrave-Macmillan, Basingstoke, UK.

Newman, P. (2004). *Creating Viable Centres: How much development is needed to create viable centres?* International Cities and Towns Society Conference. May, Fremantle.

Newman, P.W.G., Thorpe, A., Greive, S. & Armstrong, R. (2003). Locational Advantage And Disadvantage In Public Housing, Rent Assistance And Housing Loan Assistance In Perth. Final Report. AHURI Project 80038. Melbourne.

Queensland Government. (2003). *SEQ 2021.* Queensland Government, Brisbane.

State of the Environment Advisory Council. (1996). *State of the Environment Australia.* Commonwealth of Australia and CSIRO Publishing, Collingwood.

Tranter, P. (1993). *Children's Mobility in Canberra: Confinement or Independence.* Monograph Series No 7. University College, Australian Defence Force Academy, UNSW, Canberra.

Victorian Government. (2003). *Melbourne 2030.* Department of Sustainability and Environment, Melbourne.

Warren Centre. (2003). *Sustainable Transport in Sustainable Cities.* Sydney University.

Western Australian Government. (2003). *Hope for the Future: A Vision for Quality Life in Western Australia.* (Prime authors: Peter Newman and Michael Rowe). Department of the Premier and Cabinet, Perth.

10 – Sustainable work: the issues for Australia

John Burgess and Julia Connell

Australia has experienced two decades of dynamic economic reform that has included deregulation, privatisation, labour market and tax reforms. These policies have resulted in societal inequalities with almost half of the workforce now employed in jobs that are casual, part-time and/or on fixed contracts. Consequently, the contemporary workforce is divided into two groups: those in high skilled, stressful jobs who would like to work less hours, and those who have to support themselves and their families with insecure incomes. The rewards from economic growth are also very unevenly distributed. As a result, there is a need to rethink and re-conceptualise work in Australia, which has been given a narrow meaning, largely connected to market activity for the purposes of welfare policy design. Outside of the market there is much work that is neither recognised nor rewarded.

Over the past decade or so Australia's workforce has undergone enormous change. Both large and small organisations in the private, public and voluntary sectors have been subjected to environmental pressures and forces of change that have led to major transformations in organisational work structures and contracts. The responses to these influences have led to change within the industries in which Australians work, the occupations they undertake and the employment contracts they hold. Consequently, this topic underpins a number of important theoretical and policy questions such as:

- the nature and availability of employment for current and future generations
- the characteristics of future jobs and workplaces, and

- the impact of the current trajectories regarding globalisation and technology on the sustainability of work.

This chapter discusses factors related to the future of work in Australia. It outlines what has happened to work in Australia over the past 25 years and reviews recent developments concerning the workforce and the changing conditions and rewards from work, before considering the implications of these developments and the future of sustainable employment.

First, it is worthwhile defining what is meant by sustainable development in the context of employment, as so many different definitions exist. A widely used international definition is 'development which meets the needs of the present without compromising the ability of future generations to meet their own needs' (SWRA 2003). Some choose to describe these principles as 'people, pounds and planet', reflecting that sustainable development is all about achieving a balance between social, economic and environmental considerations in any decision.

Australia and the world of work – what has happened over the past 25 years?

Australia has experienced some 20 years of dynamic economic reform that has included deregulation, privatisation, labour market and tax reforms. Many economic reforms have been focused on increasing shareholder value. As such, these strategies have been translated into actions in the workplace such as downsizing, increased work intensity and unpaid overtime. The aim of creating an internationally competitive economy has worked well for business in terms of opening up markets, increasing productivity and the creation of a more 'flexible' workforce. For workers, however, there is evidence that such policies have resulted in greater societal inequalities, as almost half of the workforce is now employed in jobs that are casual, part-time and/or on fixed contracts. Statistics also indicate that over one million people are unemployed, underemployed or in 'hidden' unemployment, while others find that work itself has intensified in relation to working hours and unpaid overtime (Watson *et al.* 2003). So, this raises the question: is the possibility of economic progress and sustainable work a contradictory goal?

What we have seen is that growing richer, especially in an information era, is possible for companies without growing bigger. This means that increased national wealth does not guarantee more jobs. In fact, it can mean that a nation's corporations could grow smaller in combination with strategies such as downsizing and the drive for efficiency.

A seven-year study undertaken by Pusey (2003) investigated what 'middle Australia' (people who are neither rich nor poor) have experienced as a result of recent economic reforms and found that the vast majority believed that big business has too much power and should be more closely regulated. Corporations were perceived as the only winners from economic reform, especially their CEOs who have been rewarded with increasingly larger salary packages. Pusey concluded that Australia needs to consider international evidence showing that societies that seek to make the economy serve the people tend to be more effective (measured in terms of conventional economic indicators) than those that try to make the people serve the economy.

So in broad terms there have been both positive and negative perspectives on the future of work. Pessimists predict a divided society where jobs as such have disappeared for good (Bridges 1995) and there is mass unemployment, growing insecurity and widening social divisions. Capelli (1999) argues that the 'end of the career' will occur due to factors such as competitive pressures, volatile markets, more demanding shareholders, the ongoing need for flexibility (cost reductions), weaker trade unions and changing skill requirements. Conversely, Jacoby (1999) argues that this thesis is *not* supported by labour market evidence and the continuing experience of long-term employment in many public and private sector industries indicates that the long-term career is far from over. On the positive side, it has also been claimed that the 'new' economy will liberate many employees from dull, dreary and degrading jobs. In general, there is a lack of systematic evidence to support many of the claims on either side although much of the data presented in this chapter does tend to confirm the more pessimistic viewpoint concerning the future of work.

Recent developments concerning the Australian workforce

Although this chapter primarily focuses on the future of paid work, unpaid work is recognised as an important contributor to the economy and will be referred to later. In relation to paid work in Australia, it is suggested that the following factors are the most significant.

Increased female participation in the workforce

There has been a significant rise in female participation in paid work. Conversely, the proportion of adult men in the paid workforce is slowly declining, in common with most OECD countries. In the 1960s almost all men of working age were employed. This proportion fell to approximately 76 per cent in 2001 (OECD 2002) while the employment of women in OECD countries moved in the opposite direction. The employment rate in Australia for prime-age women (25–54 years) is now slightly higher than the OECD average for the total OECD, standing at 68 per cent in 2001 (OECD 2002: 313–15).

Persistently high rates of unemployment and underemployment

Many thousands of service workers work short hours because their industry is organised through short shifts. This frequently leads to the undertaking of multiple jobs in order to survive. Currently, one in seven workers in Australia are underemployed. Moreover, Australia has one of the highest rates of underemployment and precarious employment in the OECD, in addition to significant numbers of long-term unemployed (Watson *et al.* 2003).

Work intensity and work/life balance

All occupations and industries report that workloads have increased and that work has become more intense, with 21 per cent of people in the workforce working 50 or more hours per week. Fifty per cent of those who work overtime do not get paid for it – understaffing and work intensity have become workplace fixtures in the early 21st century. Increased workloads make it increasingly difficult to achieve work/life balance, with a large number of people in high-skilled, stressful jobs who would like to work less (Watson *et al.* (2003).

The dominance of the services sector in job creation

In common with most developed countries the service sector accounts for more than three-quarters of the economy's output and for four out of every five jobs in Australia. Many of these jobs, however, are part-time or casual and are of low quality in terms of pay, working conditions, job security and a lack of any career path (see previous point). Table 10.1 illustrates the rise in service sector employment and the decline in manufacturing employment over the period 1966–2002.

The growth in non-standard employment arrangements

This growth is particularly high in relation to part-time and casual work. New jobs created between 1985 and 2001 numbered 2.5 million but most were in industries that are characterised by low paid, part-time and casual

Table 10.1 Employment by industry, Australia, 1966, 1986 and 2002 (%)

Industry	1966	1986	2002
Manufacturing	26	16	12
Services	65	76	83

Source: Adapted from ABS (6203.0).

work – for example in the service sectors, hospitality and caring professions. Figure 10.1 illustrates that the number of full-time permanent jobs in Australia fell by 51 000 between 1990 and 2000. Three-quarters of all additional jobs created in the 1990s were part-time jobs and nearly half were part-time, casual jobs (Borland, Gregory & Sheehan 2001). Predictions are that by 2010 one in three workers will be casually employed – we are coming close to that already. Yet 68 per cent of casual workers said in a recent survey that they would prefer permanent to casual work (Watson *et al.* 2003).

Consequently, we suggest that the increase in casual work is the *major threat* to equality in the workforce and a sustainable future for Australia. These changes to employment contracts are important, as the average part-time casual job attracted earnings that were only 30 per cent of the average full-time permanent job in 2000 (and this does not include the difference in employee benefits). Casual workers are not only ineligible for holiday and sick pay benefits, they are also more vulnerable with regard to irregular income, may have to work unpredictable hours, have a lack of access to education and training opportunities in the workplace and experience job insecurity among other things (Standing 2002).

Earning inequities

There has been a huge increase in earnings at the top end of the labour market (53 per cent real income growth for those in the top decile) and no real income growth for the 60 per cent of workers who are on middle and low incomes. The working poor are no longer confined to young or part-time workers as 70 per cent of low wage workers are of prime working age (25–54 years) and the majority of low waged women work full-time (Watson *et al.* 2003).

So what can we make of these changes? Structural change has always been present as the patterns of demand and trade vary, new products emerge, consumer tastes alter and technology develops. What appears to be

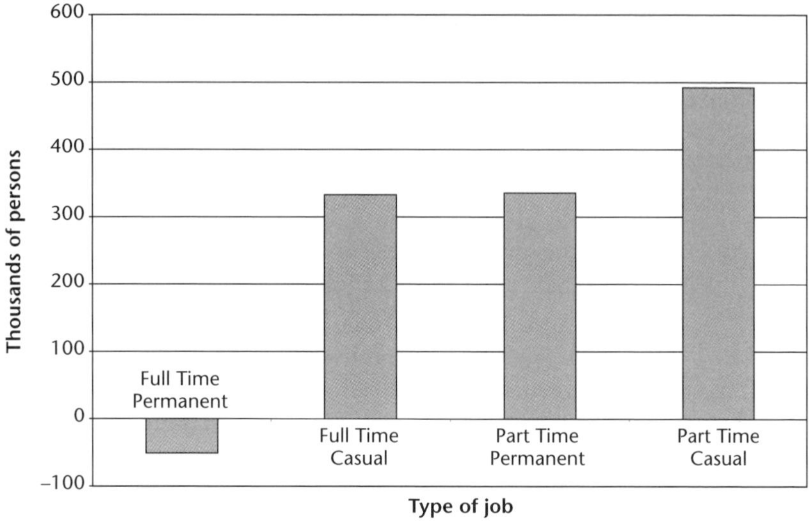

Figure 10.1 Composition of job growth, 1990 to 200 (change in thousand persons).
Source: Borland, Gregory and Sheehan (2001).

new about the current developments is the growing perception of employment insecurity, the collapse of large organisations (e.g. HIH Insurance, Ansett, OneTel), the disappearance of a job for life, the increasing ambiguity surrounding the legal status of many employment arrangements and the expectation that job content and hours are less predictable or controllable than in the past. Also, there are now many more workers who have to integrate work into other activities, especially education and caring activities. Workers appear to be confronted with uncertainty over jobs (tenure, content, control, hours and so on), have less recourse to collective representation and collective action, and are under relentless pressure to adapt and be more productive (ACTU 2003). Despite sustained growth in the economy over the past decade, unemployment and underemployment persist and the rewards from growth are very unevenly distributed across the workforce. Table 10.2 outlines some of the changes to the institutions associated with work and the rewards and conditions of work that have occurred over the past decade.

In general, Table 10.2 indicates that in Australia we have a situation where employees are working harder than ever as a result of management

Table 10.2 Changes to the institutions, rewards and conditions associated with work.

Changes to the institutions associated with work
• Shift away from centralised and industry wage determination towards enterprise and workplace wage determination
• Diminished role of industrial tribunals in the Australian industrial relations system
• Declining proportion of employees who belong to trade unions
• Decline in direct industrial activity
• Growth in ambiguous and unprotected forms of employment
• Restructuring of employment and conditions across the public sector through privatisation and outsourcing
Changes to the rewards/conditions of work
• Growth in real average full-time earnings (i.e. earnings adjusted for inflation and therefore reflecting 'real' purchasing power)
• Narrowing of the earnings differential between women and men
• Expansion after 1990 in contributors to employment linked superannuation funds
• Growth in number of employees who do not receive standard employment benefits
• Systematic shift in income distribution from labour to capital (leading to growth in profits and national income)
• Ongoing effects of income tax bracket creep on take-home pay
• Growing inequality in the distribution of earnings across the workforce
• Restructuring of the normal working week, especially since the early 1990s

practices that have been designed to address the demands of intensified competition. Moreover, changes in contemporary organisational structures have resulted in flattened career paths, whereby it is difficult to reward success and hard work through promotion.

Employers, institutional shareholders and government are clearly retreating from taking responsibility for work-related issues, resulting in the risks and costs associated with employment falling on the weakest party in a work situation (Watson *et al.* 2003). Gonos (1997) suggests that with regard to casual work, its advantage is that it offers user firms access to labour without obligation, allowing them to utilise labour while avoiding many of the specific social, legal and contractual obligations that are generally attached to employer status.

Will these trajectories continue?

Governments cannot and should not try to 'wind the clock back' on the economic change that has driven the transformation of the workforce. However, they can ensure that government policies and workplace laws provide employees with the rights, protection, information and opportunities required to succeed in the 21st century (ACTU 2003). In organisational terms, sustainability means a company will exist for more than 100 years. This requires a focus on long-term goals and giving something back to the community. Ware (cited in Kistner 2004) argues that this requires investment in educational, social and public service institutions in addition to renegotiated employee contracts that focus on work/life balance and co-investment in the future such as the development of 'talent pools'. Such initiatives are not overly evident in today's workforce in Australia. Below we have selected six major factors that workplace commentators agree appear to be continuing on a trajectory of change:

1 Gender composition

The gender composition of the workforce is likely to continue to change in the face of growing female participation and the expectation that many women have of combining career and family.

2 Service sector jobs

The majority of new jobs will be located in the service sector. Jobs *will* remain in the mining, rural, manufacturing, utility and construction sectors, but overall, their share will continue to decline. We predict a post-industrial economy where a range of caring, routine, professional, supporting and leisure-based services will dominate employment. We expect job growth to remain strong in retailing, accommodation, community services, health, education, business services and personal services. These changes contribute to the expectation that employees use their 'brains rather than their brawn' and possess superior 'soft/interpersonal skills'.

3 Non-standard work contracts

Working arrangements will continue to be fragmented, ambiguous and in many cases not regulated (consider, for example, the position of contractors and temporary agency workers). Pressures for shareholder profit and improved corporate performance will continue to lead to more innovative and flexible employment arrangements and the rewards from work will continue to be unevenly distributed. It follows that contingent or non-standard employment arrangements will become the norm. These arrangements offer flexibility for employers and choice for those who wish to combine work with study or caring responsibilities. The numbers of people holding more than one job is likely to expand. Careers will no longer mean one job with one employer in one location. Over the course of a working life, individuals can expect a career to involve many jobs, many employers, many locations and a range of occupations and skills. Such people are currently known as 'portfolio workers' (a term first coined by Charles Handy, 1989) and hold multiple jobs or contracts in multiple fields with multiple companies – they are independents with a portfolio of skills ready to be contracted out to clients on a job-to-job basis.

4 Location of work

Work will continue to spread outside of the traditional boundaries imposed by time and space such as location and set working hours. Advancing technology (such as cell phones, modems and laptop computers) support home-working, telecommuting and 24-hour employment contact. As is already occurring in some instances, the workplace will shift from city centres and head offices to homes and into cyberspace. Some work will be continuous and linked across countries and time zones. The notion of a standard working day and working week will be increasingly challenged and more people are likely to become 'portfolio workers'.

5 Unemployment and underemployment

Unemployment and underemployment are unlikely to disappear for the fundamental reasons that full employment is no longer an outcome expected from our economic system. Curiously, in the present electoral climate, politicians are not prepared to tolerate inflation but they will tolerate unemployment. Economic management is no longer assigned the responsibility for reducing unemployment since unemployment is no longer presented as a collective responsibility but as an individual responsibility. Individuals are unemployed since they do not possess the 'right' skills, the 'right' employment record, the 'right' personal characteristics or the 'right' attitude. In addition, while work remains conceptualised and constructed around the market, then there will always be those who for various reasons will be excluded from the market sector.

6 Pay inequities/work–life balance

As Watson *et al.* (2003) comment, the current system does not fairly reward effort; instead it unfairly rewards market power. This has been evident with the unrestrained earnings growth at the top end of the labour market. Unless the widening inequalities in wages are curtailed, social solidarity is unlikely to be achieved and, therefore, for those who are underemployed, receive low pay, or who have casual contracts it is implausible that they will achieve a reasonable work/life balance. This may be because they are trying to hold down two jobs to achieve a reasonable working wage or working very long hours in one job to try and retain them.

Implications for the future of work?

Polarisation and workplace inequality is likely to intensify as employment regulations become more difficult to enforce and the diversity in rewards for highly skilled and low skilled workers increase. Those who are mobile, highly skilled and adaptable can take advantage of the opportunities offered by the global labour market. Those who are not will be tied to the limited opportunities offered by local labour markets. Table 10.2 illustrates that trade union activity is declining while unprotected forms of employment are growing. Accordingly trade unions will have to rethink their organising, mobilising and servicing strategies in the face of more fragmented and insecure work arrangements. As a British trade union official commented recently, unions need to represent the small groups, all the independents who are outside the organisation and who desperately need an association to provide a range of ancillary services, such as education, legal help, protection, and advice. In addition, businesses will have to consider how to arrange their operations and labour in the context of global production, extensive outsourcing and sub-contracting possibilities and the restructuring of work through time and space.

What are the fundamental challenges facing policy makers and the community regarding the future of work? We believe there is a need to rethink and re-conceptualise work. In Australia it has been given a narrow meaning, largely connected to market activity for the purposes of welfare policy design. Consequently, worth and status have been accorded too much weight. Outside of the market there is an ongoing and significant amount of work occurring that is frequently not officially recognised nor rewarded. This needs to change, as without this type of work our communities and economy would not be able to function. Moreover, Broom (2003) argues that although these are supposedly post-feminist times, gender is still inadequately addressed in considerations of work. She contends that in the absence of gender, there is little interrogation of the relevance of market and non-market work on each other, the salience of unpaid work to men, or of the economic significance of market work to women.

For many people work gives meaning to their life. The social importance of charitable and household work reaches far beyond its economic importance as this type of work enriches family and community life, conserves cultural traditions and fosters human development (Greenwatch 1997).

Careers of the future are likely to be fragmented, disjointed, unpredictable and associated with lifelong learning and training. Work will be increasingly global and take place across borders. In addition, governments have to think about taxation and welfare systems. Where employment status is ambiguous and more workers are located outside traditional workplaces (even outside the country), the sustainability of the traditional tax base becomes questionable. Moreover, if your employment arrangements are fragmented and discontinuous, then it is difficult to develop any sustainable retirement income arrangements. One major problem for current superannuation arrangements, even without equity market bubbles, is that they are premised on a regular and sustainable full-time employment arrangement – something that is not universal in the current economy.

Towards a sustainable future of work?

Much of the economic change of the past 25 years has been driven by increased market competition. Whereas previously the quality of a firm's product or service, price and customer service were major factors influencing a firm's competitiveness in the national and international marketplace, contemporary firms are finding that their reputation for social responsibility may also influence investment decisions. The Federal Minister for the Environment and Heritage, Hon. Dr David Kemp (2002) gave an address to the Asia Society Forum where he stated that:

> ***… the pursuit of sustainability is not about ending economic growth or returning to the practices of smaller and simpler societies. It is about mobilising our intellectual and technological resources to better understand the consequences of our actions, so that we can replace unsustainable practices with sustainable ones.***

Strategies to support sustainable work practices are, however, *unlikely* to evolve at the individual firm level. Instead, we argue that coordinated government regulation is required to address the current inadequacies that are evident within the contemporary working population. Watson *et al.* (2003) argue that governments have over recent years promoted market regulation rather than institutional regulation and that it is this focus that has created many of the problems within the current labour market.

The Sustainable Europe Research Institute (SERI 2002) proposes that in order to influence political debate, principles and recommendations are needed explaining how environment and employment policies can be integrated so that they can lead to positive synergies. The emphasis on employment policies that are an integral part of sustainable development (and vice versa) works towards the achievement of joint employment and environment policy goals.

Commitment to sustainability is in the interests of not only current but also future generations. In order to look forward to a future where some form of work is an option for all, the government does have choices concerning the type of work being performed and how it is performed. Accordingly, we suggest that:

- public funding and taxation arrangements affect the allocation of resources and the composition of jobs and should, therefore, be reconsidered in the light of how they can support more sustainable and equitable work provision than currently exists
- jobs can be generated in the non-market sector (such as volunteer work) to support the unemployed and the underemployed, particularly as service sector jobs are generally labour intensive and comprise low energy use so there is no need for supporting technology
- more creative policies be explored for ways of supporting jobs through linking them to environmental sustainability. For example, taxes on polluting activities could generate revenue to assist in the financing of non-polluting activities
- long-term investments in education, training, research and public infrastructure (e.g. transport, health) be seen as capital, not current, expenditure
- qualifications in the 'eco-efficient' technologies could be promoted. If industry compliance was essential then this should be supported
- employers be encouraged to investigate job-share arrangements and shorter working weeks

- the Australian Government look to directives such as the recently published directive from the European Union on Working Conditions for Temporary Workers (Storrie 2002) to improve working conditions for casual workers in Australia. This directive intends that temporary workers will no longer be subject to discrimination due to their employment contract. Consequently temporary workers in EU member states will have the right not to be treated less favourably than comparable permanent employees.

These are just some of the suggestions that may make it possible to support both the economy and the labour force in relation to the multiple challenges of globalisation and sustainability.

In summary, of course it is impossible to predict the future. However, as the UK study on the future of work suggests (TUC Congress 2000), if something has not happened over the past 20 years, either in the UK or overseas, then we need to ask why it should happen over the next 20 years. Although it is evident that we live in a post-industrial age where the nature of work and careers is rapidly changing, we argue for long-term strategies because the social and economic consequences of changes in the world of work require careful, ongoing scrutiny by academics, employers and governments if more equitable and shared work options are to be available for the Australian majority of the future. While Australia may never see a situation where full employment is a prospect, the current state of affairs where there is growing inequality in the workplace is not sustainable either. Hence, urgent attention is required to redress the issues discussed throughout this chapter in relation to fairness, equity and choice. In other words, we need to epitomise the culture that Australia is famous for – giving a fair go to all.

References

Australian Bureau of Statistics (ABS). (2002). *Labour Force Australia.* Catalogue No. 6203.0, ABS, Canberra.

Australian Council of Trade Unions (ACTU). (2003). *Future of Work: Fairness in the Workplace.* http://www.actu.asn.au/public/futurework/workfairness.html

Borland, J., Gregory, B. & Sheehan, P. (Eds). (2001). *Work Rich, Work Poor: Technology and Economic Change in Australia.* Centre for Strategic Economic Studies, Victoria University, Melbourne.

Bridges, W. (1995). *Jobshift: How to Prosper in a Workplace without Jobs.* Allen & Unwin, St Leonards, NSW.

Broom, D. (2003). In Search of Sustainability Conference. ANU, 14 November 2003.

Cappelli, P. (1999). Career jobs are dead. *California Management Review* **42(1):** 146–67.

Gonos, G. (1997). The contest over 'employer' status in the postwar United States: The case of temporary help firms. *Law and Society Review* **31(1):** 81–110.
Greenwatch. (1997). *Germanwatch Symposium 1997: Social Sustainability*, <www.germanwatch.org/ezsymp97/work.htm> (accessed January 2004).
Handy, C. (1989). *The Age of Unreason*. Harvard Business School Press, Boston, MASS.
Jacoby, S.M. (1999). Career jobs: A debate – 'Are career jobs headed for extinction?' Career jobs are dead, Peter Cappelli; 'Reply: Premature reports of a demise'. *California Management Review* **42(1):** 123–79.
Kemp, D. (2002). Australia and Asia: Working towards sustainability speech. <www.deh.gov.au/minister/env/2002/sp25nov02.html> (accessed January 2004).
Kistner, T. (2004). *Divining the Future of Work*, Network World http://www.nwfusion.com/net.worker/news/2004/0621netlead.html
OECD. (2002). *Employment Outlook*. OECD, Paris.
Pusey, M. (2003). *The Experience of Middle Australia: The Dark Side of Economic Reform.* Cambridge University Press, Port Melbourne.
South West Regional Development (SWRA). (2003). *About Sustainable Development*, <www.southwest-ra.gov.uk/swra/ourwork/sustainable_development/index.shtml> (accessed January 2004).
Standing, G. (2002). *Beyond the New Paternalism: Basic Security as Equality.* Vero, London.
Storrie, D. (2002). *Temporary Agency Work in the European Union.* European Foundation for the Improvement of Living and Working Conditions, Dublin: <www.eurofound.eu.int> (accessed December 2002).
Sustainable Europe Research Institute (SERI). (2002). *The Future of Work and Ecology.* <www.seriat/Data/projects/aundoe/aundoemain.htm> (accessed January 2004).
Trades Union Congress (TUC). (2000). *The Future of Work, Looking Ahead – the Next Ten Years.* <www.tuc.org.uk/em_research/tuc-2397-f0.cfm> (accessed October 2003).
Watson, I., Buchanan, J., Campbell, I. & Briggs, C. (2003). *Fragmented Futures, New Challenges in Working Life.* The Federation Press, Sydney.

11 – Population – the great multiplier

Jenny Goldie

The size of the human population is central to all thinking about sustainability, though for a variety of reasons it is often ignored. The global population grows by 72 million a year and will continue at that rate for 30 to 40 years. On present projections, Australia's population in 2050 will be 27 million and the world population about 8 or 9 billion. Yet in 2004 the Earth is believed already to be 25 per cent over the limit of its regenerative and absorptive capacity. Each of the topics considered in this book is profoundly dependent on the size of the human population. There is an urgent need for both national and global population policies that take into account that the total impact of a population is equal to the average impact per capita multiplied by the number of people. Increasing population increases our impact.

Population is often overlooked in the sustainability debate. Yet it affects all the issues in the preceding chapters: water, health, land use, energy, peace, economic systems, climate, work and transport. It is the great multiplier.

Until the advent of agriculture population numbers were kept low by the local availability of food, the spread of disease and risks associated with the hunter–gatherer lifestyle (such as large carnivores). Human impact on the biosphere was minimal except where they contributed to the extinction of large mammals and birds on some continents and islands (Flannery 2001). Then with the development of agriculture, the human population grew. Subsistence agriculture had a carrying capacity – though at a higher level – defined as the number of people that could be fed by peasant farmers using manual labour, horse or oxen-ploughs, animal and human dung, and by hand-weeding of crops (Stanton 2003). Sometimes subsistence agriculture became unsustainable, for instance, when populations grew too large as on Easter Island, or irrigated lands were affected by salinisation and became unproductive as in Sumeria, and civilisations duly collapsed.

With the new technologies of the Industrial Revolution 250 years ago, however, a new carrying capacity was established. Food supplies increased dramatically and global population 'exploded', increasing over tenfold in the ensuing two and a half centuries, from 600 million to over six billion. This was also aided by better medical care that meant lower death rates from disease. In Australia, numbers grew from under half a million in c.1750 – all of whom were Aboriginal hunter-gatherers – to 20 million in late 2003. While Australia's population density is 'only' 2.5 people per square kilometre, most of the continent is arid or semi-arid and only 6 per cent is arable (Stanton 2003).

The global population continues to grow by 72.5 million a year (IPC 2004). Even with lower than replacement fertility rates in most industrialised countries, populations continue to grow owing to the demographic lag-time associated with the life spans of overlapping multiple generations. In Australia, fertility rates have been of the order of 1.7 children per woman for a number of years, yet natural increase (births minus deaths) remains well over 100 000 annually (ABS 2004). This is falling slowly, but on current trends will remain positive for another 30 to 40 years even without immigration. Despite this, net migration (immigration minus emigration) remains high and is expected to even exceed the 131 000 annually that it reached in 2002–03. By mid-century, on current trends, Australia's population will be around 27 million.

However, in the developing world, population numbers continue to explode. Almost all of the growth in the coming half-century – over two billion people – will occur in non-industrialised, poorer cities. Given that over 800 million people are already chronically malnourished, that 1.1 billion people lack clean water and over 2.4 billion proper sanitation, this further growth is potentially catastrophic (Postel 2004).

A study by Mathis Wackernagel and his team, published by the US National Academy of Sciences, found that the world passed its regenerative and absorptive capacity back in 1979 and with each subsequent year the deficit is increased by 1 per cent (Wackernagel *et al.* 2002). This means that in 2004 we are 25 per cent over the limit – we are satisfying our excessive demands by consuming the Earth's natural assets. We are now consuming natural capital as well as interest, that is, we are not living sustainably. We have exceeded the new carrying capacity.

Humans' impact on the biosphere is more than a function of mere numbers. Over three decades ago, Paul Ehrlich and John Holdren introduced the I=PAT formula (Ehrlich & Holdren 1974). The environmental impact of a population (I) is the product of the size of the population (P), its affluence (A – a measure of consumption) and the technology required (T) to bring material goods to the population. Because T is difficult to quantify, a simpler way to consider total impact is that it is equivalent to the product of average individual impact and the number of people in a population. So even if per capita impact stayed the same, a doubling of the population would double the total impact. While this formula is from an environmental perspective, it also has implications for social scientists and economists. For instance, assuming standards of living stay about the same, a doubling of the population will require a doubling of infrastructure and services. Those genuinely concerned about sustainability cannot ignore this multiplier effect.

We now discuss the other issues addressed in this book and how population affects them.

Water

A sustainable and secure society is one that meets its water needs without destroying the ecosystems upon which it depends or the prospects of generations to come (Postel & Vickers 2004). Freshwater is, unfortunately, distributed unevenly between and within countries and many nations are already water-stressed, that is, unable to meet their water needs in a sustainable way. While over a billion people do not have ready access to potable water, this figure may rise to as much as five billion in 25 years because of both global warming and population growth. This is likely to trigger conflict across Asia and Africa (McGuire 2002).

In Australia, tensions already exist between States over the waters of the Murray–Darling Basin. Though (as Peter Cullen notes in Chapter 6), most States have moved beyond simplistic water planning that focuses only on extracting water for irrigation. They now understand that rivers must have water left in them for ecological purposes. Extractions from the Murray–Darling Basin were capped at 1994 levels in an effort to arrest the severe deterioration in the river's health. But the cap was set too high, for even returning a further 1500 GL to the river annually gives only a moderate chance of a healthy river, according to Cullen. The economic development of the Murray–Darling Basin is limited by the availability of water, though there is room for greater efficiencies and a move from low to high value crops.

Any further population growth in the Basin will mean some water must be transferred to domestic users from irrigators, who are still fighting the withdrawal of water for environmental purposes. Nevertheless, most people now recognise that water is a finite resource that must somehow be divided fairly between the river itself, agriculture, industry and domestic use. It is dawning on the planners and legislators of Canberra, the national capital sited within the Basin, that if the city's population is to grow to a half million from its current 320 000 then permanent water restrictions will be necessary. Further dams to ensure water security for Canberra's citizens, while technically and economically possible, will probably be deemed ecologically unacceptable because they will restrict environmental flows downstream.

Meanwhile coastal cities, also afflicted by water restrictions, continue to grow unabated. As Cullen notes, Melbourne has insufficient water in the existing catchment for the anticipated extra million people. On top of improved efficiency, other measures such as the recycling of water will have to be employed to ensure adequate supply. Infrastructure for wastewater treatment or a third pipeline for recycled water all cost money, however, and the 'balancing of the economic, environmental and social costs of such decisions is not readily achieved' as Cullen notes. At some point, further population growth must be deemed unaffordable and indeed, unsustainable.

Some believe that, despite water being a finite resource, improvements in efficiency will allow Australia's population to grow for the foreseeable future. Certainly, there have been considerable innovations in agriculture such as drip irrigation for horticultural crops that have not only saved water but impeded salinisation by reducing leakage beyond the root zone. As far as domestic use is concerned, however, it is generally agreed among water utilities that a saving of only 25 per cent can be expected (Perkins 2002). Given the uncertainties of climate change and the extent of overextraction, caution is warranted in terms of further population growth.

Health

Tony McMichael (Chapter 2) reminds us that 'humankind, via its expansions in numbers and economic intensity, is now overloading the biosphere' and that the resurgence of infectious diseases has 're-emphasised population-level phenomena'. Historically, the rate of spread of communicable diseases has wholly depended on the growth and concentration of human populations. The unprecedented population densities in 14th century Europe led to the plague outbreak that wiped out a quarter of the population.

Around 40 per cent of people now live in cities of over a million people, a function of rural–urban migration and population growth. Overcrowding, or the increased proximity of susceptible individuals, favours the spread of infectious diseases such as tuberculosis, dengue fever, malaria and acute respiratory illnesses. In addition, the kind of rapid urbanisation as seen in developing countries is often characterised by a lack of clean water and poor sanitation, leading to water-borne diseases such as cholera and diarrhoea. Disease vectors, such as mosquitoes and rats, thrive in such cities, further facilitating the spread of disease (Brown *et al.* 1999).

Degradation of the global environment as a result of population numbers quadrupling and economic activity increasing twentyfold over the past century, however, has the gravest implications for health. As McMichael notes, it has changed the conditions of life on Earth 'by altering global climate, depleting stratospheric ozone, extinguishing whole species and their local populations and damaging food producing ecosystems on land and at sea'. The effects are felt mostly at local and national levels.

Global climate change has contributed to an increase in natural disasters over the last three decades of the 20th century. One billion or so people suffered from them, mainly in poorer countries (McGuire 2002). Developing countries are particularly susceptible to disasters such as mudslides, cyclones and flooding because rapid population growth has meant that people increasingly utilise high risk terrains such as steep hillsides, flood-plains and coastal zones for habitation and farming. Many developing countries are also in the tropics and these areas experience higher rates of natural disasters than temperate countries.

Land use and natural ecosystems

John Williams and Denis Saunders (Chapter 5) note that current land use in Australia is unsustainable. It is characterised by extensive loss of species, changes in ecosystem processes and increasing degradation of land and water resources. Much of the degradation is a result of extensive land clearing and replacement of long-rooted native perennials with short-rooted annuals. This causes watertables to rise, bringing salts to the surface.

Many argue that population has little to do with this as much of the clearing occurred in the early days of European settlement when there were fewer than four million people living in Australia. In fact, as much clearing has occurred in the past 50 years as in the first 150 years of settlement. In 2000, over half a million hectares were cleared, mainly in Queensland and New South Wales (ABS 2003). Land clearing has been primarily for agriculture, the products of which have gone to feed and clothe an expanding domestic population and also millions of people overseas. While agriculture is not directly related to the size of our domestic population, it is to global population. It is related in an indirect way because it pays for imports to maintain Australia's standard of living.

More recently, as people along the east, south-east and south-west coasts of the continent have spilled out of the major cities, there has been extensive land clearing for urban development. Indeed, the greatest growth has been along the coast. Regrettably, these are often regions of high biodiversity, notably south-east Queensland, where the habitat of that great Australian icon, the koala, is now threatened. Human population density in Australia tends to be highest in the regions containing the greatest number of species so while agriculture – driven by increased consumption and population growth – remains the major cause of biodiversity loss in this country, urban development resulting from population spread is increasingly implicated.

Indeed, the *Australia: State of the Environment Report 2001* (Australian SoE Committee 2001) was explicit in stating:

> ***Population growth has particular effects on coastal Australia. Urban sprawl, high energy consumption, stormwater pollution of estuaries and coastal waters, and the continued decline of biodiversity as a result of land-clearing all arise from population and economic pressures.***

And the Australian Bureau of Statistics (ABS 2003) corroborated this, noting:

> ***A significant factor causing pressure on some parts of Australia's coastline is high population density in coastal regions, particularly along the east and south-east coasts and along the west coast south of Perth … The coastal strip is an ecologically sensitive zone, and urban***

sprawl, pollution of rivers, lakes and seas, were described by the Resource Assessment Commission as the two most important problems faced by the coastal zone.

Given the limitations on inland rivers to support large conurbations of people, further population growth is likely to be on the coast which, without dramatic improvement in management, is likely to see further deterioration of coastal habitats and marine ecosystems.

Energy

The State of the Environment Report (SOE 2001) notes that Australia's energy use is very high by world standards and continues to grow rapidly. It reports that with the associated emissions in greenhouse gases, energy use is 'increasingly at an unsustainable level'. Sustainable energy sources are having an effect in reducing emissions 'but are not enough to offset the impact of increasing demand'.

Between 1990–91 and 1998–99, Australia's total energy consumption rose 23 per cent while population rose by just under 10 per cent (ABS 2003). While the rise in consumption exceeded population growth, the latter is a significant contributor to demand and often negates any gains in energy efficiency.

It is important to look at where the energy is used. In 1996–97, 56 per cent of Australia's energy-related greenhouse gases were emitted in the production and consumption of goods and services for households, mainly through electricity use and motor vehicles (ABS 2003). The combination of population growth (increasing) and the number of people per household (declining) means the number of households is climbing inexorably and consequently so is energy use (Liu *et al.* 2003).

Andrew Blakers (Chapter 8) argues that we can move to a solar economy by mid-century provided funds for research and development are directed appropriately. But while so much of Australia's energy comes from fossil fuels and more than half goes to households, a brake on population growth appears warranted.

Economic systems

Clive Hamilton and Richard Denniss (Chapter 4) argue that instead of pursuing economic growth, we now need to seek a balance between the economy, the well-being of the community and the sustainability of the natural environment. A lower economic growth rate would reduce the demand for labour that, in the short term, could increase the unemployment rate. But in the longer term, 'stabilising population size will reduce the pressure to create the new jobs required each year to absorb new entrants into the labour market'.

In the recent public debate over ageing, the major concern has supposedly been a lack of workers in the future when there will be an excess of dependents for each taxpayer (the dependency ratio). The debate was hyped-up by the business and housing lobbies – those with a vested interest in promoting population growth. It largely ignored the option of importing workers when needed from an overcrowded world, overlooked the possibilities of increasing the participation and productivity rates, and appeared ignorant of the fact that Australia's dependency ratio was quite healthy and would remain so for another 20 years. Nevertheless, on the assumption that economic growth would continue at current rates, the concern about lack of workers in 30–40 years time did have some validity.

If, however, our society does move to a low-growth or zero-growth scenario, the concern about supply of future workers loses its urgency since fewer workers will be needed. Of course, very low fertility rates (below 1.5 children per woman) will distort the age structure unduly and are best avoided. On the other hand, the need for ecological sustainability may require that we reduce population numbers as soon as possible. In order to achieve this we must maintain fertility above 1.5 but below 2.1 and reduce immigration (which, due to the age of most immigrants, has only a marginal effect on averting an ageing population).

For too long, the debate on population has been characterised by inconsistency and irrationality. In Tasmania the cry is for more migrants. Yet the logging industry is ravaging old-growth forests and when environmentalists complain, the response is: 'We need the jobs'. It might have some credibility were there not such a simultaneous campaign to increase the number of workers.

Inequality and conflict

Colin Butler (Chapter 3) cites resource scarcity and inequality as fundamental drivers of human conflict. Related to these are population and environment. In his now-famous 1994 essay 'The Coming Anarchy', Robert D.

Kaplan described how anarchy follows overpopulation, resource depletion and environmental decline (Kaplan 1994). 'It is time to understand "the environment" for what it is: the national security issue of the early twenty-first century,' Kaplan wrote. He used the West African nations with which he was familiar as examples of countries descending into anarchy.

Later in 1994, the central African country of Rwanda collapsed into genocidal civil war, leaving a million dead. Another two million fled the country. A struggle between the rival Tutsi and Hutu elites for control of the Rwandan state was the chief cause (Homer-Dixon 1999). Nevertheless, land scarcity was a contributing factor. Most people were agricultural peasants and the nation had the highest population density in Africa. The growth rate was 3.4 per cent, that is, population doubling time was a mere 21 years. (Stanton 2003).

In other parts of the world where population growth rates and rural densities are high, land scarcity exacerbates tensions between groups. In the Indian state of Bihar, for instance, it has deepended divisions between landholding and peasant castes, bringing land reform to a halt. Similarly in Haiti, shortages of forests and soil have inflamed tensions between social groups, duly obstructing technical and institutional reform (Homer-Dixon 1999).

It is commonly said that the wars of the 21st century will be fought over another resource – water. Water is closely tied to the production of food. Of the more than 300 river systems that cross national boundaries, those that have the most potential for conflict run through those populous countries where damming by upstream nations will impede the ability of downstream neighbours to produce adequate food for ever-growing populations, notably: the Nile (Ethiopia, Sudan, Egypt); Ganges (Pakistan, India, Bangladesh); Mekong (China, Cambodia, Laos, Vietnam); and the Tigris–Euphrates (Turkey, Syria, Iraq).

Climate

Graeme Pearman argues (Chapter 7) that climate change is the first truly global challenge for sustainability. To stabilise concentrations of greenhouse gas emissions will require cuts of 70 per cent or more in current global emissions. Yet today two billion people do not have access to the levels of energy that we enjoy, and approximately another two billion people will be added to the world by mid-century. Poverty alleviation, however, requires the availability of the amenity of energy. In Pearman's view, the two requirements of emission reduction and energy amenity are contradictory.

The only hope in resolving the dilemma is a rapid move to renewable energy and away from a fossil fuel-based economy as well as a full-scale effort to limit population growth in the most humane, non-coercive ways possible. Achievement of the necessary reductions in emissions would be significantly easier without an extra two billion consumers. Even though these consumers will use less energy per capita than had they been born in the developed world, the added emissions will be significant. They may be forced to use 'dirty' energy like brown coal, or cheap energy such as firewood, because it will be a while before they can benefit from technologically advanced, energy efficient systems currently being developed.

We in the developed world must curtail our use of fossil fuel. Business-as-usual is simply untenable since climate change from even a two degree Centigrade increase will wreak havoc. A five or six degree increase will see the likelihood of a Permian-style extinction that saw 95 per cent of species obliterated from the face of the Earth. There can be no other course but to reduce emissions significantly and stabilise the atmosphere. Continued population growth, whether here or in the developing world, works against achieving this critical objective.

Work

John Burgess and Julia Connell (Chapter 10) report that over one million people in Australia are currently unemployed, underemployed or in hidden employment. This could be addressed by taking some of the work away from those who are currently over-worked through shorter working weeks, job-sharing or other measures. But it also suggests that there are too many people or workers for the jobs currently available. This is likely to change with ageing of the population when there may be insufficient workers, but it is not the case now despite a few specific areas of skills shortage.

Nevertheless, the Howard Government has already adopted an aggressive stance towards importing workers rather than waiting until they are really needed. It has significantly increased skilled immigration (another 5000 on top of the current 61 000 in April 2004) although the humanitarian stream (around 12 000) remains almost untouched (ABS 2004). 'Skills' include everything from plasterers and hairdressers to neurosurgeons. Most could and indeed should be trained in Australia, were sufficient resources diverted to our educational institutions. At present many immigrants come from developing countries which can ill afford to lose their own skilled workers. This is nothing short of poaching.

While some fill gaps in the workforce, and a few genuinely create jobs, many compete with existing Australian workers, some of whom are new immigrants themselves. There is not only competition for jobs but also competition for housing. Birrell and Healy (2003) noted that the decline in housing affordability in Australia has coincided with a sharp rise in immigration. Because of its role in household formation, immigration affects demand. Housing prices are shaped by other factors, such as the financial environment and the capacity of the building industry to meet demand. Nevertheless by 2021, according to Birrell and Healy's projections, the migration component of household formation growth in Sydney will be around 75 per cent, in Melbourne and Adelaide 60 per cent, and in Perth 54 per cent.

The higher cost of housing, resulting in part from higher demand due to population growth, explains why many workers are working longer hours or second jobs (when they can get them) and inevitably leads to hardship amongst the under and unemployed.

Urban design and transport

Peter Newman (Chapter 9) argues for urban growth boundaries as an essential feature for transforming a car-dependent city into one where sustainable transport has some possibility of succeeding. Given the imminence of the oil crisis (production will peak and demand will then exceed supply) and the urgent need to reduce greenhouse gas emissions by over 70 per cent, it is imperative that we move away from our dependence on private cars and redesign urban areas appropriately.

The concept of discrete cities with boundaries is an admirable one for a range of environmental, social and economic reasons. Nevertheless, if population growth continues so will household growth. At some point a new city will need to be created, paving over yet more countryside and emitting ever more greenhouse gases in the building of houses, schools, hospitals and workplaces, even when those buildings adopt all the best features of ecological innovation with respect to energy, water, site ecology, building materials, and waste management.

Similarly, the construction of buses and trains entails considerable use of materials and energy. Until they can be run purely on renewable energy (such as electricity from wind power) they too will contribute to greenhouse gas emissions. Even bicycles, powered ultimately by food consumed by humans, require materials and energy to manufacture.

So while all efficiency initiatives in the area of transport and urban design are to be applauded, they must go hand in hand with a cap on population growth if sustainability is to be achieved.

Conclusion

Twenty-five years ago, in 1979, the world passed its regenerative and absorptive capacity, that is, exceeded its human carrying capacity. The cause was a combination of an explosion in human numbers and their economic activities. Focus in the sustainability debate has too often just been on economic activities – reducing consumption, moving to a renewable energy economy and so on – while the population aspects have been ignored. Numbers are stabilising in the industrialised world. However, many nations in the developing world are still experiencing very high and absolutely unsustainable rates of growth. In Australia, growth continues despite lower than replacement fertility rates through a combination of natural increase and high net overseas migration.

Population growth often negates efficiency measures. For instance, many Australian cities are implementing ever more stringent water-efficiency measures but continuing population growth offsets these. Energy-efficient appliances are being promoted, quite rightly, but if the overall number of consumers increases then little is gained with respect to sustainability.

A combination of resource scarcity, environmental decline and overpopulation can lead to anarchy and other forms of conflict. Water shortages – the result of global warming and too many people for the water available – are looming as a major source of conflict in the 21st century.

The greatest challenge to sustainability is global climate change, a function of too many people and their activities. Unless greenhouse emissions are reduced 70 per cent or more and the atmosphere stabilised, global warming could wreak havoc on the Earth. Yet we must do this while lifting an existing two billion out of poverty, and coping with another two billion or more people in the next half century. Moving from a fossil fuel-based economy to a renewable one based on solar power is essential, along with sharing energy more equitably. These efforts, however, must go with stabilising and even reducing population numbers until we are back within the carrying capacity of the Earth.

Australia was a signatory to the Plan of Action arising out of the United Nations International Conference on Population and Development 1994. We are obliged to contribute generously to family planning activities in developing countries and to other programs that help bring down the birthrate, notably education for girls and primary health care. Australia's contribution, however, has fallen far short of its promises. Overseas aid is still way below the UN-recommended 0.7 per cent of GDP, and the family planning program within that is even more inadequate.

Australia faces a number of ethical dilemmas with respect to population. It has an obligation to keep its population low for the sake of preserving natural habitat for future generations. Yet by contributing to global warming through excess greenhouse gas emissions, it should take in at least some of those displaced by rising sea levels. To what extent it should take in large numbers of economic refugees from countries who have failed to curb their own numbers, however, is a matter for further public debate.

References

Australian Bureau of Statistics (ABS). (2004). Australian Demographic Statistics, March 2004. Publication no. 3101.0

Australian Bureau of Statistics (ABS). (2003). *Environment by Numbers – Selected Articles on Australia's Environment.* ABS Catalogue no. 4617.0. Commonwealth of Australia.

Australian State of the Environment Committee (SoE). (2001). *Australia State of the Environment 2001.* Independent Report to the Commonwealth Minister for the Environment and Heritage. CSIRO Publishing on behalf of the Department of the Environment and Heritage, Canberra.

Birrell, B. & Healy, E. (2003). Migration and the housing affordability crisis. *People and Place.* Vol 11. No. 3. Centre for Population and Urban Research, Monash University, Melbourne.

Brown, L., Gardner, G. & Halweil, B. (eds) (1999). Infectious disease. In *Beyond Malthus: Nineteen Dimensions of the Population Challenge.* pp. 57–60. WW Norton, New York.

Ehrlich, P. & Holdren, J. (1974). Impact of population growth. *Science* **171:** 1212–17.

Flannery, T. (2001). *The Eternal Frontier: An Ecological History of North America and its Peoples.* Atlantic Monthly Press, New York.

Homer-Dixon, T.F (1999). *Environment, Scarcity and Violence.* Princeton University Press, Princeton and Oxford.

IPC. (2004). <http://www.census.gov/cgi-bi/ipc/population>

Kaplan, R.D. (1994). The coming anarchy. *The Atlantic Monthly* **273(2):** 44–76.

Liu, J., Daily, G., Ehrlich, P. & Luck, G. (2003). Effects of household dynamics on resource consumption and biodiversity. *Nature* **421:** 530–33.

McGuire, B. (2002). *A Guide to the End of the World. Everything You Never Wanted to Know.* Oxford University Press, Oxford.

Perkins, P. (2002). Answer to question, public meeting, Canberra, April 2002.

Postel, S. & Vickers, A. (2004). Boosting water productivity. In *State of the World 2004: Progress Towards a Sustainable Society.* Chapter 3. (Ed. Worldwatch Institute.) Earthscan, UK.

Stanton, W. (2003). *The Rapid Growth of Human Population 1750–2000.* Multi-Science Publishing Company Ltd, UK.

Wackernagel, M., Schulz, N.B., Deumling, D., Linares, A.C., Jenkins, M., Kapos, V., Monfreda, C., Loh, J., Myers, N., Norgaard, R. & Randers, J. (2002). Tracking the ecological overshoot of the human economy. *Proceedings of the National Academy of Sciences* **99(14):** 9266–71.

12 – Achieving a sustainable future

Ian Lowe

The scale and seriousness of Australia's social and environmental problems are no longer in doubt. We are consuming the opportunities of future Australian generations by our lifestyle choices. And at the global level, human activities are altering the very systems on which our species depends. Doing nothing about the many problems identified in this book, such as climate change, is not an option. It is possible to move to a sustainable future but it will require fundamental changes to our values and national social institutions. We must recognise that we share a common future with the entire human family and the other species with which we share the planet. It is not just our humanitarian duty to improve the lot of the poorest people of the world. It is also enlightened self-interest because a world of increasing inequality will be a world of increasingly destructive tension.

The evidence is clear. If civilisation is to survive, the next century will have to be a time of transformation – not just in technological capacity but also in our approach to the natural world, and to each other. The second report in the United Nations Environment Program (UNEP) series on the global environmental outlook, *GEO2000*, said: 'The present approach is not sustainable. Doing nothing is no longer an option.' A sustainable society would not be eroding its resource base, causing serious environmental damage or producing unacceptable social problems. It is clear from the earlier chapters that our present lifestyle does not satisfy *any* of these main criteria. We are dissipating resources future generations will need, damaging environmental systems, and reducing social stability by widening the gap between rich and poor. In so doing, we are acting irresponsibly. It is possible to move to a sustainable future, but it requires fundamental changes to our values and social institutions. There is hope in that human systems can change radically very quickly. There is growing global recognition of the need for change. The serious obstacle is the dominant mind-set of decision-makers who don't recognise the problem, or see the possible solutions as threatening their short-term interests. While UNEP said that doing nothing about the huge problems we face is not an option, it remains the most common response of decision-makers today.

Reconciliation

As a first and fundamental issue, we have no prospect of being a sustainable society until we recognise the injustice done by our forebears to the Indigenous people of Australia. Their land was acquired by a process that can only be described as robbery with violence, their lifestyles and cultures have been systematically destroyed and their living conditions today shame us as a civilised nation. We cannot restore what we have taken, but we must acknowledge our debt. We should work with representatives of Indigenous Australian peoples to develop institutions that will allow us to share this country and build a sustainable future.

Resources

A sustainable society will not erode its resource base. Our most serious medium-term resource problem is oil. Our entire transport system is based around petroleum fuels that are cheaper than any other liquid except tap water. We pay more per litre for beer, cask wine, milk, orange juice and even bottled water than we do for petrol. Optimists think the peak of world oil production may be 15 years away, while the pessimists believe it was in the year 2000. However, the age of plentiful cheap petroleum fuels will end soon. There are technical alternatives, most obviously fuel cells using hydrogen produced by splitting water with solar electricity, but they are at present expensive. We should be planning now for the post-petroleum age – as the forward-looking oil companies are.

There are other resource issues. Humans now use about half of all the world's available fresh water directly or indirectly, but 1.2 billion people do not have clean drinking water. Our forests, our fisheries, our agricultural soils and our grazing lands have not been used sustainably. We have to reduce resource use so that we use the income of natural systems; our present approach is running down the capital. We must improve our scientific knowledge so we know sustainable use levels. In the case of Australia's major inland river system, the Murray–Darling, present water usage is unsustainable. Governments must agree on firm targets for environmental flows based on scientific understanding of the needs of the river system. We may move to a system of tradable water rights, as this is an economically efficient way to use a fixed amount of water. That approach is morally justifiable only if we know that the overall extraction rate is ecologically sustainable; otherwise we are producing short-term economic benefits at the expense of future generations.

Economic issues

Two hundred years ago, the economic trend was acquiring global empires from which to obtain cheap resources and open up markets. A hundred years ago, the fashion was protection within each nation. Fifty years ago, it was public funding of infrastructure projects by government. Today's fad is a return to 18th century economic empires, once again allowing free rein to global movement of capital.

In local terms, Australia is already what Barry Jones called a 'post-industrial economy'. The traditional sectors of agriculture, mining and manufacturing now account for less than a quarter of the economy and fewer than 20 per cent of jobs. Today most of us work in the 'services' sector. It is increasingly hard for Australia to compete in global markets as politicians reduce protection for local produce. Our relatively high wages and salaries make our production costs high in labour-intensive fields like textiles, clothing and footwear. Our low level of investment in research and innovation makes us uncompetitive in value-added manufactures. With few import restrictions, we consume overseas-produced goods at an ever-increasing rate, so we have to keep expanding the production of low-value commodities from our rural land. The only economic bright spots are our attractiveness as a tourist destination and the demand for coastal housing, driven by an increasing population, an ageing society and an increasing leisure preference. The proposed 'free' trade agreement with the USA has serious costs; it appears likely to accelerate the decline of Australian value-added manufactures and services, as well as undermining our cultural institutions.

A sustainable economic future must be based on reward structures that direct our limited resources into areas in which we have some prospect of being competitive. We also need to recognise that the scale of economic activity is not a measure of community well-being and move on to develop indicators of genuine progress. It is indefensible to persist in the naïve assumption that economic growth will produce social and environmental benefits, given the solid statistical evidence that the unprecedented economic performance since 1990 has coincided with limited social benefits and serious environmental decline (ABS 2002).

Social cohesion

Fifty years ago, Australia was one of the most equal societies in the world. The legal framework of the basic wage set a minimum standard, while the highest salaries were only five to ten times that minimum. A strong commitment to public education gave opportunities to bright children from a range of disadvantaged backgrounds. Most people used public transport, while the health system made provision on the basis of need. Today, we are one of the most unequal of all the industrialised nations. Executives are given salary packages hundreds of times the average household income, often with arrangements for tax avoidance built in. Public health care and education have been steadily run down, and Commonwealth ministers tell those who miss out on hospital beds or university places that they should be paying for private options. This is cruel advice to the increasing fraction of the adult workforce that is unemployed, underemployed or doing poorly paid casual work. Many Australian households today have great difficulty affording the basic needs of food, power and clothing. Low-income families do not have the luxury of buying educational opportunities or privileged health care. If we continue rationing life opportunities according to income, we must develop policies that systematically reduce the inequality of household incomes.

While the links are not simple, the widening divisions between rich and poor are inevitably causing resentment and social tensions. While regions can do little about national policies that bring about an unequal society, local communities can take actions to improve public transport, public education and community services. This is a wise investment in future social cohesion. Children growing up anywhere in Australia should have the opportunity to realise their potential in all ways – physically, intellectually and emotionally.

We should also be responsible global citizens. It is not just our humanitarian duty to try to improve the lot of the poorest people of the world. It is also enlightened self-interest because a world of increasing inequality will be a world of increasing tension. We cannot be secure doing property deals on mobile telephones from four-wheel-drive vehicles in a world where most people have never ridden in a car, made a telephone call or owned any property. We should take a much more generous attitude to aid, lift our commitment to the international target of 0.7 per cent of Gross Domestic Product, and cancel the debts of those countries for which debt repayments constitute a grinding financial burden. As the Australian delegation said at the 1999 UNESCO World Conference on Science, 'We should aim to make this not just a new century but a **just** new century.'

Environmental issues

The scale and seriousness of environmental problems are no longer in doubt. At the national level, two *State of the Environment* reports have now been published. The 1996 report showed that we have a beautiful and unique environment, with many aspects in good condition by international standards, but we also have very serious problems, most obviously loss of biological diversity, degradation of inland waterways and destruction of the productive capacity of rural land. Its final section linked the environmental problems to lifestyle choices, showing that a sustainable future requires integrating environmental awareness into all social and economic decisions. The second report, released in 2002, noted an improvement in urban air quality but found that all the other critical environmental problems are getting worse, because of the increasing ***pressures*** on natural systems. Each year the Australian population grows by about 200 000, as the 'natural increase' (the excess of births over deaths) of about 120 000 is augmented by migration. The material expectations of people also increase each year. We use more energy, travel further in larger and less efficient cars, live in larger houses, consume more resources and produce more waste.

The compounding effect of more people, is putting ever-increasing pressure on our natural systems. The Australian Bureau of Statistics confirmed the decline in their report *Measuring Australia's Progress*. It analysed the decade 1990–2000. All the usual economic indicators showed positive trends. The social indicators were mixed, with some very serious negative trends. Of all the environmental indicators discussed, only urban air quality improved. The report showed more land being cleared, more species threatened, declining river health, more degraded land and increasing greenhouse gas emissions. The obvious conclusion is that the increasing economic production from the natural systems of Australia is at an environmental cost. Tim Flannery made this point about the unsustainable use of Australia's natural resources in *The Future Eaters*, arguing that we are consuming the opportunities of future generations by our lifestyle choices (1994).

Global studies draw the same conclusion. UNEP has now produced three reports in its *Global Environmental Outlook* series. They show some successes, such as the concerted international effort to stop releasing the chemicals that deplete the ozone layer and 'encouraging reductions in many countries' of urban air pollution. They also document 'environmental challenges' – increasing emissions of greenhouse gases, over-exploitation of water, 1200 million people without clean drinking water and twice that number without sanitation, increasing numbers of species being lost, fisheries in decline, land degradation, and problems caused by increasing release of nitrogen into natural systems.

A new report by the International Geosphere-Biosphere Programme, *Global Change and the Earth System: A Planet Under Pressure,* paints a disturbing picture. It says human activities are affecting global systems 'in complex, interactive and apparently accelerating ways', so that we now have the capacity to alter natural systems in ways 'that threaten the very processes and components ... on which the human species depends'. The most urgent issue is climate change. The politicians of the developed world accepted the scientific arguments when they negotiated the Kyoto agreement to slow down release of greenhouse gases. Only the Bush regime in Washington and the Howard Government in Canberra have refused to ratify this treaty.

The Earth as a whole has warmed about 0.6 degrees Centigrade in the last hundred years, with Australia warming slightly more than the global average. The Earth is now warmer than at any time since credible records began. As predicted by climate scientists, there have been other changes associated with the warming: shrinking of glaciers, thinning of polar ice, rising sea levels, changing rainfall patterns and more frequent extreme events like droughts and severe storms. While we only have solid information about the southern hemisphere for two hundred years, the record for the northern hemisphere suggests it is now warmer than at any time in the last two thousand years.

The world body of scientific expertise, the Inter-governmental Panel on Climate Change (IPCC), has refined its mathematical models of climate and compared the models with records. They now take into account the different possible future patterns of fuel use as well as the uncertainties in the science. The most optimistic view is that the temperature this century will increase by another 1.4 degrees. That low forecast assumes that we will rapidly phase out fossil fuels and hopes scientific uncertainties will work in our favour. A pessimistic view gives an increase of about 2.4 degrees, even with a concerted global campaign to reduce fuel use. More realistic assumptions about fuel use lead to estimated temperature increases between 2 degrees and about 4.5, while the extreme inaction advocated by some could cause increases as high as about 6 degrees.

Climate change is having many effects, not all of them negative. For example, some plants have improved growth because of the combined impact of more carbon dioxide, higher temperatures and changing rainfall patterns. But the changes are having a wide range of effects on crops and on natural systems. It is not sensible to ignore the problem. Just as other serious environmental problems have been tackled at the international level, we need a concerted global response to climate change, not a 'head in the sand' approach. We should set a target of substantially reducing fossil fuel use, say to 30 per cent of present levels by 2050. Cutting our emissions is our obligation to the world community. It will require policy measures from all levels of government to encourage cleaner energy supply and much more efficient conversion of energy into the services we need. The obvious way to fund the transition is to systematically phase out the huge current subsidies of fossil fuel supply and use (NIEIR 1996), transferring those public funds to the expansion of renewable energy supply technologies and efficiency gains.

In structural terms, we should undertake systematic improvement of energy standards for domestic and commercial buildings as well as vehicles and appliances. Urban planning should emphasise making services accessible. At present it seems to concentrate on mobility, a feature needed only when essential services are not readily accessible. Fuel use for transport has to be reduced. Instead of allowing or even encouraging large four-wheel-drive vehicles in urban areas, they should be phased out. There must be incentives for small efficient vehicles, public transport, cycling and walking. Solar hot water should be mandatory in buildings throughout Australia.

Human health and well-being

It is important that we recognise the link between healthy ecological systems and human health. At its most basic, we rely on natural systems to provide the essentials of life: oxygen, water and food. We also need those systems to process our wastes. Some systemic features that contribute to poor health result directly from the degradation of the natural world's capacity to provide those essential services. More broadly, nature provides our sense of place, our cultural identity and spiritual sustenance. Again, we are healthier and more fulfilled when these needs are satisfied. An investment in the health of our natural systems is also an investment in the health of the community.

A sustainable future?

So how can we achieve a sustainable future? At a political level we certainly need to move beyond the simplistic view that economic growth will solve our problems. As Clive Hamilton showed in his recent book *The Growth Fetish*, the ABS indicators were no surprise (2003). In societies like ours where most people have all the essentials of a decent life and more, economic growth does not make people happier or more fulfilled, but comes at social and environmental costs. The 1996 *State of the Environment* report said that successful remedies for environmental problems took a comprehensive and systematic approach, whereas failures were usually piecemeal efforts that attacked symptoms rather than underlying causes. Since some of the underlying causes are linked to lifestyle choices, the systematic approach must include those aspects of lifestyle.

In *Resetting the Compass*, Yencken and Wilkinson suggest a guide for 'Australia's Journey Towards Sustainability' (2000). Existing policies will not achieve a transition to sustainability because they do not address the growing pressures of increasing population and rising material demands per person. So, they conclude, we need to aim to stabilise the population as well as commit ourselves to 'dematerialisation'. A German study argued that Europe needs to reduce energy use by a factor of four and materials use by a factor of ten. Several European nations have adopted those targets! More generally, we must recognise that sustainability has many dimensions. We have to accept that growth has costs as well as benefits, and we should be more concerned about the *quality* of growth than the rate.

The vision of a sustainable future involves using sustainable resources, a commitment to maintaining the ecological values of natural systems, development of social cohesion, nurturing our cultural traditions and finding durable economic activities. Balancing these is a complex task that defies simple approaches. We have to make difficult decisions about things that must change and those that should be preserved. This will be possible only if there is an open and transparent process, involving the community and allowing time to work through the costs and benefits of alternatives. Changing one thing in a complex system always produces other changes, so no change is ever universally beneficial; there are always losers as well as winners. In a fair world, those who lose out from a change that benefits the community as a whole should be compensated by the rest of the community. That principle was accepted when we decided that those using Sydney airport should pay a noise levy to compensate those under the flight path. It is a good general principle.

Finally, moving toward a sustainable future will require new institutions. Concerted action is frustrated by division of responsibility between different levels of government, by State and local boundaries that have no social or ecological logic, and by the silos of established departments and traditional disciplines. We need new structures that will enable coordinated policies and actions that integrate environmental goals with our social, economic and cultural aspirations.

Broader issues

At the international level, there is a growing awareness that a sustainable future will involve significant change. *Our Common Future* states that the world's economic and environmental futures are intertwined and should be seen as complementary, rather than in competition. The UNEP report *GEO-3* set out some of the principles for change by exploring four possible scenarios. In *Markets First*, globalisation and a liberal trade agenda promote rapid economic growth, but the cost is increasing environmental damage. In *Security First*, the wealthy use force to suppress growing protest against ecological problems and a widening gap between rich and poor, creating a divided and violent world. In *Policy First*, governments take decisive action to curb environmental excesses, but it proves difficult to bring the material living standards of the poorer countries up to an acceptable level.

The most hopeful scenario, *Sustainability First*, is based on a shift in values, allowing us to reach a global consensus on satisfaction of basic needs for all within the limits of natural systems. Couching the problem in these terms clearly shows that the present world is a long way from having the *values* needed for the transition to sustainability. We also lack the knowledge to be confident that we are interacting sustainably with natural systems. Great changes can be made, in principle, by policy reform to reduce the resource demands and environmental consequences of our lifestyle. The problem is that the political will to implement such a strategy is nowhere in sight. As *Great Transition* says, policy reform has to overcome 'the resistance of special interests, the myopia of narrow outlooks and the inertia of complacency'. As long as politicians are more concerned about the next election than the next generation, necessary reforms won't happen.

Great Transition argues that market-led wealth generation and government-led technological change need to be supplemented and guided by a values-led move to an alternative global vision, based around such principles as equity. We should see the economy as a means of serving our needs within the limits of natural systems, rather than an end in itself. We need new technologies based on renewable resources, efficient use and 'industrial ecology' – that is seeing the waste of one industrial process as the feedstock of another. It is possible to eliminate hunger by stabilising the Earth's population and improving distribution systems. Above all, we should aspire to *genuine* globalisation, recognising that we share a common future with the entire human family and all other species.

This vision is utopian, but that has been said of all important reform movements. Those who opposed slavery just over two hundred years ago were told that no economy could function without slave labour, while the suffragettes were persecuted when they sought the vote for women a hundred years ago. Closer to our time, 20 years ago it was still considered utopian to dream of Berlin without the Wall, or South Africa without apartheid. Most social reforms we now take for granted were initially denounced as unachievable. They happened because determined people worked for a better world. All around the globe, people are striving to develop social and institutional responses that will enable the transition to a sustainable future. A naïve faith in markets will not bring it about. We have to change our values, recognising that we share the Earth with all other species and hold it in trust for all future generations. That is our moral duty to those future generations, our own descendants.

Acknowledgement

This chapter has been significantly improved by Patricia Kelly's helpful criticism and suggestions.

References

Australian Bureau of Statistics. (2002). *Measuring Australia's Progress.* ABS Cat. No. 1370.0. Commonwealth of Australia, Canberra.

Australian State of the Environment Committee. (2001). *Australia State of the Environment 2001*. Independent Report to the Commonwealth Minister for the Environment and Heritage. CSIRO Publishing on behalf of the Department of the Environment and Heritage, Canberra.

Council of Australian Governments. (1992). *National Strategy for Ecologically Sustainable Development*. Commonwealth of Australia, Canberra.

Flannery, T. (1994). *The Future Eaters*. Reed Books, Chatswood.

Global Scenarios Group. (2002). *Great Transition*. Stockholm Environment Institute, Boston.

Hamilton, C. (2003). *The Growth Fetish*. Allen & Unwin, Sydney.

National Institute of Economic and Industry Research. (1996). *Subsidies to the use of Natural Resources*. Environmental Economics Research Paper No. 2. Department of Environment, Sport and Territories, Canberra.

Raskin, P., Banuri, T., Gallopín, G., Gutman, P., Hammond, A., Kates, R. & Swart, R. *(2002). Great Transition: The Promise and Lure of the Times Ahead.* A report of the Global Scenario Group. Stockholm Environment Institute, Boston.

State of the Environment Council. (1996) *State of the Environment Australia 1996*. Commonwealth of Australia and CSIRO Publishing, Collingwood.

Steffen, W., Jager, J., Matson, P., Moore, B., Oldfield, F., Richardson, K., Sanderson, A., Schnellnhuber, J., Turner, B.L., Tyson, P. & Wasson, R. (2004). *Global Change and the Earth System: A Planet Under Pressure*. Springer-Verlag, Berlin.

United Nations Environment Program. (1999). *Global Environmental Outlook 2000*. Earthscan, London.

United Nations Environment Program. (2002). *Global Environmental Outlook 3*. Earthscan, London.

World Commission on Environment and Development. (1987). *Our Common Future*. Oxford University Press, Oxford.

Yencken, D. & Wilkinson, D. (2000). *Resetting the Compass*. CSIRO Publishing, Collingwood.

Appendix – Recommendations of the ISOS conferences

This appendix contains selected extracts and all of the 33 recommendations that were contained in the communiqué from the 2003 ISOS ('In Search Of Sustainability') internet and face-to-face conferences co-hosted by Australia 21, Nature and Society Forum and Sustainable Population Australia. The full communiqué can be downloaded from the conference website: <www.isosconference.org.au>.

Australians are not living sustainably and we are part of a shrinking and interconnected world that is becoming increasingly hostile to the long-term survival of our species.

In moving towards a sustainable future, we need to ensure that we:

- avoid eroding the natural resource base
- provide substantial global equality of opportunity
- have some sort of cultural and spiritual foundation for the future
- keep the size of the human population within ecological limits.

In spite of growing recognition and action by Australian Governments and industry in these matters, our society still falls far short of meeting these criteria.

The consensus emerging from the internet discussions which have been taking place during the past nine months between Australian experts and lay-people is that we must now, as a nation, make significant changes in the way we manage our social, economic and environmental assets.

The necessary changes need to be coordinated and integrated at all levels of government. Industrial and community bodies also need to be involved in the changes that will alter the way we govern and manage our lives. Our progress towards a more just and sustainable global society must be understood and monitored by all Australian citizens.

Material progress has been the dominant paradigm of modern Western societies like Australia. We have tended to view progress as a pipeline: pump more wealth in one end, and more welfare flows out the other. Economic growth in this outdated model is paramount.

Sustainable development, which must now supplant this previously dominant paradigm, does not accord economic growth overriding priority. It seeks a better balance and integration of social, environmental and economic goals and objectives to produce a high, equitable and lasting quality of life. Sustainable development acknowledges the dynamic relationship between the goals of improving well-being and ensuring that improvements are compatible with a healthy natural environment.

A sustainable future global society is technically possible, economically feasible and socially desirable, but we need the political will to manage the transition. Our science gives us better understanding than ever before of the natural world and our impacts on it. Technology gives us unprecedented capacity to change the world to meet our needs and suit our desires. Humanity requires us to use that scientific understanding and technological capacity to develop as a sustainable society. That is a moral responsibility to future generations as well as to the other species with which we share the planet.

Recommendations

1 Water

1.1 That Federal and State Governments, as a matter of priority, each develop a discussion paper on optimal use of our limited water supply.

1.2 That State Governments immediately establish a transparent framework for monitoring the short and long-term social and environmental impact of water reform.

1.3 That the Murray–Darling Commission establish firm targets and timelines for the return of environmental flows to the Murray–Darling River system and for other rivers from which water has been over-extracted. The relevant States should then implement a program.

1.4 That State Governments set and implement strong mandatory standards for the efficiency of showerheads, taps and other water-using equipment.

2 Human health and well-being

2.1 That universities recognise the complexity of our social and environmental systems and then apply systems science, within an interdisciplinary framework, to the relationship between human population health and the sustainability of the natural and social environments.

2.2 That education syllabuses assist children to better understand that, in addition to achieving a fair and cohesive social environment, they must live within the limits of the natural world. This will require a rediscovery of our connections to the natural world which some may see as a new form of spirituality. Children also need to understand the importance of nutrition, exercise patterns and social interactions to their own health.

2.3 That we move from away from the competitiveness, selfishness and short-termism of nation-states (a modern analogue of ancestral warring tribes) to a more cooperative approach so we can better shape the future.

2.4 That, in order to develop a new societal awareness, we recognise that better management of the natural environment is required as well as the way we design and manage our urban environments.

3 *Land use and natural ecosystems*

3.1 That State Governments set up a system whereby farmers cease broad-scale clearing of native vegetation to prevent further deterioration of water quality, increases in dryland salinity, further degradation of soils and further losses of native biota.

3.2 That the Federal Government legislate to implement accounting rules for all businesses and government agencies to value ecosystem services, such as production of clean water, maintenance of fresh air, conservation of healthy soils in natural resource management policy and reward those who are maintaining essential ecosystem services.

3.3 That the Federal Government set up a process to reach agreement on the magnitude of hidden environmental subsidies in the production costs of food, fibre and water, and then introduce levies equivalent to those costs to fund environmental restoration.

3.4 That Federal and State Governments adequately fund research into farming systems and land-use patterns that work with, and not against, the environment.

4 Energy

4.1 That Federal and State Governments set targets, with interim milestones, for a reduction to 30 per cent of current fossil fuel use by 2050. Clear long-term targets will minimise the cost of the transition to a sustainable energy economy.

4.2 That Federal, State and Local Governments adopt policy measures and strategies that will allow these targets to be met, including intensive energy efficiency measures and support for renewable energy.

4.3 That Federal and State Governments fund the transition to the above target by removing subsidies to the production and use of fossil fuels and redirect the funding on an interim basis to efficient energy use and renewable sources of energy.

5 Equity and peace

5.1 That the Australian Government urgently increase its commitment to international assistance to 0.7 per cent of gross domestic product, recognising that it must also reconsider the way partnerships are developed with poorer nations.

5.2 That the Australian Government cancel the debt of those developing nations where the debt represents an unsustainable financial burden.

5.3 That government and business promote, at all available international forums, fair trade as an absolute imperative in the reduction of global inequalities.

5.4 That all Australian Governments develop 'Global Studies programs' for participation by secondary school students.

6 Economic systems

6.1 That State Governments include, as part of school curricula, consideration of the societal choices we make in adopting specific elements of an economic system.

6.2 That the Federal Government invest in the development of valid long-term indicators of social and environmental progress which can be balanced against measures of the health of the economy and of economic growth.

6.3 That a formal inquiry be undertaken, along the lines of the Hilmer Enquiry into competition policy, to identify the structural mechanisms and legislative changes required to integrate social and environmental sustainability into decision making at all levels of Australian governance.

7 *Climate*

7.1 That the Australian Government fund, as a national priority, the development of the technologies and institutions for non-greenhouse-emitting energy production.

7.2 That the Australian Government invest in scientific research that improves the forecasting of weather and climate events that impact on water management, insurance, energy demands, health, and the management of agricultural and natural systems.

7.3 That, as with other elements of the sustainability agenda, media and community education agencies become better informed on the options that now confront us with respect to policy on climate change.

8 *Labour force and work*

8.1 That a national inquiry into the nature, meaning, organisation and funding of work in a sustainable society be undertaken. The inquiry should bring together a broad cross-section of thinkers and policy-makers with no constraints on the outcome.

8.2 That Australia commit to the element of the Universal Declaration on Human Rights which reads: 'Everyone has the right to work, to free choice of employment, to just and favourable conditions of work and to protection against unemployment', and develop constructive proposals to the ILO to assist its implementation globally.

9 *Transport and urban design*

9.1 That new Australian building approvals require the average house and commercial building to use 50 per cent less energy and water as well as have minimum sustainability outcomes with respect to waste, site ecology, transport and affordability.

9.2 That, in developing national transport policy, energy and CO_2 emission balance sheets for different transport options is provided so alternatives to the current road/truck-oriented transport system are encouraged.

9.3 That the Federal Government promote Australian building innovations (e.g. eco-cement) that contribute global solutions towards sustainability; provide more sustainable city innovation R&D funds; and re-direct some housing and transport funds towards Sustainable Cities demonstration projects.

10 Need for new Australian structures and institutions

The conference noted the recent call by the Australian Collaboration for the following overarching developments as we confront the sustainability 'bottleneck' of the 21st century.

10.1 A new national political charter with commitments to specific social, cultural and environmental as well as economic goals.

10.2 The establishment of a new parliamentary institution to examine long-term trends affecting Australia and to foster wide debate about them in Australian society.

10.3 The establishment of a formal independent inquiry into the working of democracy in all its forms in Australia and a commitment to implement its main recommendations.

10.4 Action to put in place comprehensive reporting of social, cultural, environmental and economic conditions and trends.

The characteristics of a sustainable society are that it is humane, has an eco-centric approach and a long time horizon, and is informed, efficient and resourced.

The communiqué concluded: '*If all Australian governments committed to building social, environmental and economic sustainability into every element of governance, we could be an example to the rest of the world.*'

Index